QUANTUM MECHANICS, CELL-CELL SIGNALING, AND EVOLUTION

QUANTUM MECHANICS, CELL-CELL SIGNALING, AND EVOLUTION

JOHN S. TORDAY
Professor of Pediatrics, Obstetrics and Gynecology, and Evolutionary Medicine, University of California, Los Angeles, Los Angeles, CA, United States

Fellow, The European Academy of Science and Arts, Salzburg, Austria

ELSEVIER

ACADEMIC PRESS
An imprint of Elsevier

Academic Press is an imprint of Elsevier
125 London Wall, London EC2Y 5AS, United Kingdom
525 B Street, Suite 1650, San Diego, CA 92101, United States
50 Hampshire Street, 5th Floor, Cambridge, MA 02139, United States
The Boulevard, Langford Lane, Kidlington, Oxford OX5 1GB, United Kingdom

ISBN: 978-0-323-91297-6

For information on all Academic Press publications visit our website at https://www.elsevier.com/books-and-journals

Publisher: Andre G. Wolff
Acquisitions Editor: Glyn Jones
Editorial Project Manager: Timothy Bennett
Production Project Manager: Punithavathy Govindaradjane
Cover Designer: Vicky Pearson

Typeset by STRAIVE, India

Dedication

I would like to dedicate this book to the people that I love, namely, my wife Barbara, my daughter Nicole, my son Daniel, my daughter-in-law Erin, and my grandchildren Abigail Eve and Delia Rose.

I would also like to dedicate this book to my mentors Claude Giroud and Mary Ellen Avery.

Contents

Preface

The consensus is that Quantum Mechanics is the foundation for all of existence, inanimate and animate alike. The problem in applying it to physiology is what is referred to as the "explanatory gap," or decoherence, which refers to the unlikelihood that there could be an interface between Quantum Mechanics and the "classical" wet, sloppy nature of physiology. However, a Bayesian approach to the evolution of physiology has determined that Symbiogenesis is actually founded on Quantum Entanglement—that is, when gravity impinges on a curved surface like that of a lipid-based micelle, it generates energy that would have been used to sustain the Quantum Entanglement of particles within protocells. With this perspective in mind, the following chapters address the application of Quantum Mechanics to the evolution of physiology from its very beginning as the transition from matter to life. No longer do we have to fear that in the way that Mary Shelley related the tale in her book *Frankenstein, the Modern Prometheus*. Now that we know the causal nature of life, we can embrace it in order to truly understand our origin and fate, and live freely without fear, or so is my hope in presenting this work to you, dear reader.

John S. Torday
Professor of Pediatrics, Obstetrics and Gynecology, and Evolutionary Medicine, University of California, Los Angeles, Los Angeles, CA, United States
Fellow, The European Academy of Science and Arts, Salzburg, Austria

Acknowledgments

I have been supported by research grants for my entire career, freeing me to do the experiments that are largely the basis for this book. Among them have been the National Heart, Lung and Blood Institute; the March of Dimes; the Thrasher Foundation; the American Heart Association; the Hood Foundation; the Parker B. Francis Foundation; The Peabody Foundation; The Meadows Foundation; the Edward Beatty Foundation; and the King Trust.

Thank you all.

CHAPTER 1

The cell membrane as a functional Mobius strip

Introduction

If you cut a Mobius strip formed by two twists, not the conventional one twist, in half, the edges will form a Trefoil Knot. That knot can be untied to form a circle, proving that it is a true mathematical knot (Kauffman, 2006). The cell is homologous with the mathematical knot since it, too, must be able to unknot itself to form the egg and sperm during meiosis in order to reproduce. The homology between the knot and the cell is thought-provoking biologically because the Trefoil Knot is representative of the endoderm, ectoderm, and mesoderm, the three germ layers of the gastrula that ultimately form the embryo, beginning with the zygote. Upon further reflection, the cell membrane functions as if it were a Mobius strip, forming one continuous surface between the inner environment of the cell and the outer environment from which it has evolved. However, it is not formed by taking a circular surface, cutting it, twisting it, and attaching the two ends as you would conventionally form a Mobius strip. Conversely, David Bohm's Explicate Order forms a boundary with the Implicate Order. That lipid boundary is the prima facie Mobius strip that divides the infinite surface of the Implicate Order into inside and outside by "recalling" its preadaptive state as individual lipid molecules before there was an inside or outside.

But all of that begs the question, "Why is there an interrelationship between a mathematical knot and a cell?". They are both circular in two dimensions and spherical in three dimensions. In the case of the cell, the sphere is the most efficient physiologic form for function because metabolically it is the optimal ratio of surface area to volume for gas exchange. In the case of the geometrical sphere, Schwarzschild's Radius is the mathematical reduction of Einstein's

Quantum Mechanics, Cell-Cell Signaling, and Evolution
https://doi.org/10.1016/B978-0-323-91297-6.00004-7

Field Theory to a Black Hole. The circular form of the lipid barrier generated by micelles forms a "hole" in the infinite plane of the Implicate Order; homologously, a Black Hole forms a hole in the Cosmos. Cosmic fibers course through the Cosmos in much the same way that the cytoskeleton confers form to the cell, functionally determining whether it is homeostatic, meiotic, or mitotic, controlled by Target of Rapamycin signaling.

The cell membrane as the prototype for the Mobius strip

Upon further consideration, the cell membrane is like a Mobius strip because it forms a continuous topologic surface between the outer and inner environments of the cell, having given rise to the "inside" and "outside" when micelles first formed. In other words, when lipids were first immersed in water, they "invented the concept" of the Mobius strip, because prior to that, there was no inside or outside—it was one infinite plane in the Implicate Order.

To further unpack this, the origin of the cell was derived from amphiphilic lipid molecules floating on the surface of the waters that covered the Earth 100 million years after its formation, the negatively charged ends pointing downward into the water and the positively charged ends pointing upward. When enough lipid molecules aligned with one another and packed together to reduce the surface tension of the water surrounding themselves, by inhibiting the Van der Waals forces that create the surface tension of water, they spontaneously formed micelles, or lipid spheres. So those micelles realized the Explicate Order as distinct from the Implicate Order, as described in David Bohm's classic book "Wholeness and the Implicate Order." The preadaptation for micelles was the lipid molecules immersed in water, but there was no inside or outside; inside and outside only appeared once the micelles emerged from those lipid molecules. Therefore, the micelle was the origin of the Mobius strip.

The hypothesis being tested is that the lipid barrier of the Explicate Order "conceived of the Mobius strip" by dividing the Implicate Order into "inside and outside," conceiving life in the process.

That was the origin of consciousness as the First Principles of Physiology, referencing the Singularity that predated the Big Bang.

Micelles, semipermeable membranes, and calcium ion fluxes

Micelle semipermeable membranes allow particles to enter and exit it due to osmotic pressure, forming the basis for the protocell. Calcium ions were prevalent in the primordial ocean, being dissolved from the bedrock. As carbon dioxide produced by plants built up in the atmosphere, it dissolved in the water to produce carbonic acid, accelerating the dissolution of the bedrock, releasing evermore calcium into the water.

The accumulation of calcium ions within the protocells was hastened by the Sun warming them by day, causing the lipid membranes to liquify and therefore to expand; at night, the micelles would cool and re-conform to their original configuration due to hysteresis, or "molecular memory." That recursive expansion and contraction caused buildup of calcium ions within the protocells. But since calcium ions are toxic to lipids, causing them to denature, as a consequence a subset of protocells hypothetically evolved the capacity to control the entry and exit of calcium ions using calcium channels.

Calcium ions passing through the cell constitute the biologic flow of energy

Experimentally, if cells are exposed to zero gravity, they lose their capacity to generate a calcium ion flow. This critical role of gravity in the formation and initiation of life even applies to abiotic micelles (Claassen and Spooner, 1996). The causal relationship between the force of gravity and the "life force" per se intimates the relationship between physics and biology. That interrelationship transcends the cell because Einstein's Field Theory states that when gravity impinges on a curved surface, it produces energy. The principle of Quantum Entanglement would have entrained particles within the micelle, stabilized by such gravity-dependent energy. Based on Quantum Mechanical principles, the configuration of such entangled particles would have referenced nonlocal events in the Cosmos.

This foundation for the micelle would hypothetically have provided the preadaptive condition for the emergence of symbiogenesis—the acquisition of factors in the environment that posed an existential threat during the "history" of the organism (Sagan, 1967), or what we commonly refer to as "cellular evolution" (Torday and Rehan, 2012).

The evolution of multicellularity

Over the course of cellular evolution, there was a point at which prokaryotes (bacteria) devised ways of mimicking multicellularity, namely Quorum Sensing and Biofilm, presenting a biologic advantage over unicellular eukaryotes. In order to compete, eukaryotes devised means of expressing true multicellularity using cell-cell communications based on soluble growth factors signaling to their cognate receptors on neighboring cells.

Beginning with the fertilized egg, the animal and vegetal poles intercommunicate using such growth factors to initiate cell division and differentiation that ultimately terminate in the offspring. Beyond birth, the same cell-cell signaling mechanisms maintain homeostasis within and between the cells that constitute the organism. Breakdown in cell-cell communication leads to remodeling of structure and function, constrained by homeostasis (Torday, 2015). Physiologic stress can cause the disruption of homeostatic control, leading to the production of reactive oxygen species (ROS); ROS can cause gene mutations and duplications (Storr et al., 2013), giving rise to a variation in cell-cell signaling. Once a new homeostatic set-point is attained, the cells involved will maintain and perpetuate that state structurally and functionally in what we conventionally recognize as evolution.

For example, the lung evolved from the swim bladder of boney fish, both of which are "gas-exchangers." In the case of the former, the swim bladder utilizes gases to facilitate efficient feeding at different depths of water, exploiting such gases to inflate or deflate the bladder for buoyancy, optimizing metabolism by minimizing energy output. In the case of the latter, the lung facilitates the uptake of oxygen for metabolic activity within the tissues and organs of the body—same genes, different purposes, or what is descriptively referred to in biology as pleiotropy.

The above narrative for the emergence of vertebrates from water to land exemplifies the "long game" of evolution. Carbon dioxide in the atmosphere built up due to the evolution of plant life on Earth (Romer, 1949). CO_2 in the atmosphere caused a "greenhouse effect," warming the ambient temperature of the atmosphere. That process had a twofold effect on life, reducing the amount of oxygen dissolved in the water, partially drying up the water covering the Earth, exposing land masses for subsequent habitation. The combined effect resulted in a subset of boney fish attempting to move out of water onto land (Clack, 2012). It was the so-called physostomous species of boney fish that formed the basis for land vertebrates, having a pneumatic duct that connects the esophagus to the swim bladder, forming that structural homology between fish and land vertebrates (Zheng et al., 2011).

Tiktaalik is fossil evidence for this water-land transition, representing the transition from fins to legs (Daeschler et al., 2006). The "fossil record" for the evolution of the internal organs is more difficult to determine since there is no actual fossil evidence. However, there are cellular-molecular data embedded in ontogeny and phylogeny as the documented history of the organism, and cell-cell signaling allows recall of such events (Torday and Rehan, 2017).

The fossil skeletal evidence indicates that vertebrates attempted to breach land on at least five separate occasions (Ashley-Ross et al., 2013), inferring a stepwise "ratcheting-up" process for evolving from the swim bladder to the lung. The final common biomolecular pathway for this series of events is lung surfactant, the "soapy" material that prevents the alveoli from collapsing (atelectasis) under the hydrostatic force of the water tension caused by the fluid lining the alveoli. Surfactant is present in the swim bladder, where it prevents the walls of the bladder from sticking together. The bladder evolved into alveoli through a progressive decrease in the surface area of the gas-exchange chamber, effectively increasing the surface-area-to-blood-volume ratio for oxygenation. But in order for that to occur, there had to have been tandem increases in the amount and/or quality of surfactant since surface tension is inversely related to the diameter of a sphere based on the Law of Laplace. As the lung evolved, both of these properties of lung surfactant have been documented to have changed commensurately to one degree or another, averring the causal interrelationships.

Over the course of this challenging process for land adaptation, other physiologic systems evolved in tandem, including the endocrine system. As fish evolved into quadrupeds, emerging from the buoyancy in water facilitated by the swim bladder to the lung for adaptation to land life, there was a concomitant increase in the effective force of gravity on the organism. As a result, there were three documented receptor gene duplications: the Parathyroid Hormone-related Protein Receptor (PTHrPR) (Pinheiro et al., 2012), the Glucocorticoid Receptor (GR) (Bridgham et al., 2006), and the β-Adrenergic Receptor (βAR) (Aris-Brosou et al., 2009). All three of these receptor gene duplications were critically important for the evolution of the lung, since they all resulted in the amplification of the signaling pathways that they determine. The PTHrPR was existential for the formation of alveoli since its deletion results in death at birth in mice; the βAR was essential for the evolution of the pulmonary circulatory system independently from the systemic circulatory system; the GR was vital for the increase in the βAR receptor density within the aforementioned lung circulatory system. The selection pressure for the PTHrPR was likely initially due to its role in the calcification of bone, strengthening the skeletal system in order to support the organism's increased weight on land, and again, selection pressure for the GR evolved from the increase in blood pressure on land due to gravity, causing shear stress on the microcirculation, which would have been offset by the evolution of the GR from the mineralocorticoid receptor, epistatically offsetting the effect of the latter on blood pressure. The advent of the GR would have facilitated the distribution and number of βARs, particularly in allowing for the independent regulation of the lung circulatory system, as mentioned above.

Epigenetic inheritance

Seen from this prima facie perspective, it is of value to consider the mechanism of epigenetic inheritance. Elsewhere, it has been hypothesized that the phenotype is not merely the physical form of the organism, but is actually the means by which it expresses "agency" in order to monitor the environment for changes that might prove life-threatening. By assimilating such changes in the egg and sperm as biochemical adducts of DNA that modify its readout, the offspring are enabled to adapt to such environmental changes.

Conclusions

In this way, the cell is able to perpetuate itself, defined by the cell membrane delineating the outside of the cell from the inside. As the origin of the Mobius strip, the cell membrane "remembers" its history, referring all the way back to those lipids produced by Pulsars in deep space, orienting themselves upward toward the Sun, conducting thermal energy to fuel life. The innate, holistic connection between the lipid origin of life and the topology of the Mobius strip attests to the Singularity of Nature (Torday, 2019). Elsewhere (Torday and Rehan, 2017), it has been postulated that the unicell is the primary state of being and that its purpose is to remain in close proximity to the Singularity. The idea that the Explicate Order formed the basis for the Mobius strip as both inside and outside is consistent with that perspective.

But seeing this phenomenon retrospectively is misleading. For example, Hoffmeyer (1998) has deduced that the cell membrane behaves like a Mobius strip, yet he makes the same error in judgment that Knot Theoreticians (Kauffman, 2006) or Rewrite Math (Rowlands, 2014) do by thinking of such processes only in the "now" synchronically, whereas the basis for these interrelationships must be seen diachronically "across space-time" in order to recognize that primacy of the flow of energy over the material epiphenomena. By way of explanation, it is like Husserl explaining the origin of geometry being innate to our being (Husserl, 1939), the geometric figures being projections of our innate knowledge of such forms, like Plato's Theory of Forms being archetypes in our minds. So, too, are mathematical "knots" archetypes, which must be unknottable as circles. Such circles are homologs of the unicell, which also knots after a fashion developmentally to form complicated organisms, but must also unknot meiotically to produce the egg and sperm. It is noteworthy that it has been discovered that there are "fibers" in the Cosmos that course through the heavens, much like the cytoskeleton, which determines the physiologic state of the cell. The cytoskeleton could hypothetically be queried as a "bioassay" for the Cosmological Fibers in much the same way that Smolin has expressed the organic nature of the Cosmos (Smolin, 1997).

And Peter Rowlands' Rewrite Math is homologous with biologic evolution, the cell being the equivalent of "zero" as an attractor, and the introduction of a new value to the Rewrite Math requiring an evaluation of the existing set of values (Rowlands, 2014). The same occurs in epigenetic evolution, new epigenetic data having to be evaluated by the genome of the egg or sperm before it is assimilated (Skinner, 2016).

In short, the above comparison and contrast between induction and deduction is differentiated by the judicious use of experimental data to leverage knowledge, moving ever closer to Bohm's Implicate Order (Bohm, 1980), but never quite reaching it since we have evolved out of ambiguity (Torday and Miller Jr., 2017) and must remain in that state as our raison d'etre. In so doing, we are not only able to understand Heisenberg "uncertainty," we live by its principle as the embodiment of the atom (Torday and Miller Jr., 2017).

References

Aris-Brosou, S., Chen, X., Perry, S.F., Moon, T.W., 2009. Timing of the functional diversification of alpha- and beta-adrenoceptors in fish and other vertebrates. Ann. N.Y. Acad. Sci. 1163, 343–347.

Ashley-Ross, M.A., Hsieh, S.T., Gibb, A.C., Blob, R.W., 2013. Vertebrate land invasions- past, present, and future: an introduction to the symposium. Integr. Comp. Biol. 53, 192–196.

Bohm, D., 1980. Wholeness and the Implicate Order. Routledge, London.

Bridgham, J.T., Carroll, S.M., Thornton, J.W., 2006. Evolution of hormone-receptor complexity by molecular exploitation. Science 312, 97–101.

Claassen, D.E., Spooner, B.S., 1996. Liposome formation in microgravity. Adv. Space Res. 17, 151–160.

Clack, J.A., 2012. Gaining Ground. Indiana University Press, Bloomington.

Daeschler, E.B., Shubin, N.H., Jenkins Jr., F.A., 2006. A Devonian tetrapod-like fish and the evolution of the tetrapod body plan. Nature 440, 757–763.

Hoffmeyer, J., 1998. Surfaces inside surfaces. On the origin of agency and life. Cyber. Hum. Knowing 5, 33–42.

Husserl, E., 1939. The origin of geometry. Rev. Int. Philos. 1, 157–179.

Kauffman, L.H., 2006. Formal Knot Theory. Dover Publications, Garden City.

Pinheiro, P.L., Cardoso, J.C., Power, D.M., Canario, A.V., 2012. Functional characterization and evolution of PTH/PTHrP receptors: insights from the chicken. BMC Evol. Biol. 12, 110.
Romer, A.S., 1949. The Vertebrate Story. University of Chicago Press, Chicago.
Rowlands, P., 2014. The Foundations of Physical Law. World Scientific, Singapore.
Sagan, L., 1967. On the origin of mitosing cells. J. Theor. Biol. 14, 255–274.
Skinner, M.K., 2016. Endocrine disruptors in 2015: epigenetic transgenerational inheritance. Nat. Rev. Endocrinol. 12, 68–70.
Smolin, L., 1997. The Life of the Cosmos. Oxford University Press, Oxford.
Storr, S.J., Woolston, C.M., Zhang, Y., Martin, S.G., 2013. Redox environment, free radical, and oxidative DNA damage. Antioxid. Redox Signal. 18, 2399–2408.
Torday, J.S., 2015. Homeostasis as the mechanism of evolution. Biology 4, 573–590.
Torday, J.S., 2019. The singularity of nature. Prog. Biophys. Mol. Biol. 142, 23–31.
Torday, J.S., Miller Jr., W.B., 2017. The resolution of ambiguity as the basis for life: a cellular bridge between Western reductionism and eastern holism. Prog. Biophys. Mol. Biol. 131, 288–297.
Torday, J.S., Rehan, V.K., 2012. Evolution, Cell-Cell Communication and Complex Disease. Wiley, Hoboken.
Torday, J.S., Rehan, V.K., 2017. Evolution, the Logic of Biology. Wiley, Hoboken.
Zheng, W., Wang, Z., Collins, J.E., Andrews, R.M., Stemple, D., Gong, Z., 2011. Comparative transcriptome analyses indicate molecular homology of zebrafish swimbladder and mammalian lung. PLoS One 6, 24019.

Further reading

Johnson, M., 2006. Molecular mechanisms of beta(2)-adrenergic receptor function, response, and regulation. J. Allergy Clin. Immunol. 117, 18–24.
Smolin, L., 1999. The Life of the Cosmos. Oxford University Press, Oxford.
Torday, J.S., 2018. Pleiotropy, the physiologic basis for biologic fields. Prog. Biophys. Mol. Biol. 136, 37–39.
Torday, J., 2021. Cellular evolution as the flow of energy. Prog. Biophys. Mol. Biol. 167, 147–151.

CHAPTER 2

Consciousness, the "missing link" between physics and physiology

Etienne Roux (2014) tells us that defining physiology based on function is teleologic and tautologic. If so, then what does physiology actually constitute? The following chapters test the hypothesis that what we think of as physiology is actually its assimilation of the Cosmos (Torday, 2020), which maps onto physiology, and as such forms the basis for what we think of as consciousness.

Currently, we are no closer to understanding what consciousness is now than we were during the era of the Ancient Greek Philosophers. There are numerous reasons for this failure, beginning with the lack of scientific evidence, compounded by the fact that we are all survivors, by definition, rendering us confident in our narcissistic, self-centered view of the world, characterized by the Anthropic Principle, that we are "in" the Cosmos, whereas in reality, we are "of" the Cosmos based on Symbiogenesis Theory, the concept put forward by Lynn Margulis that we have evolved through the assimilation of factors in our environment and have rendered them as what we think of as physiology. In a series of articles and monographs beginning in 2007, Torday and Rehan have, for the first time, recognized the value in understanding evolution as the merging of physics and physiology, allowing for the synchronic understanding of such processes juxtaposed with the diachronic "across space-time" view, which eliminates the material aspects of physiology, revealing the flow of energy as the primary process of evolution (Torday, 2021). Seen in this way, from the unicell as the origin of life reveals the mechanism behind emergence as Gould and Vrba's exaptation, or preadaptation, which are Bayesian. In this way of understanding the causal nature of serial

Quantum Mechanics, Cell-Cell Signaling, and Evolution
https://doi.org/10.1016/B978-0-323-91297-6.00009-6

organism-environment interactions, the underlying cellular-molecular mediators of structural-functional evolution can finally be understood for the first time.

Advancement in understanding consciousness has been complicated by the state of both the physical and biologic sciences as probabilistic chance occurrences, dissuading us from any attempt to understand the origin of our self-awareness. And yet ironically, if we were to determine how and why we have evolved, it would ease our minds, so in that vein the following concept will be put forward—we know now that we have evolved through the process of Symbiogenesis, assimilating factors in the environment that have posed existential threats. In turn, development recapitulates that process through the mechanism of cell-cell signaling, culminating in the homeostatically controlled mechanisms of physiology. And it has been proposed that those processes have been in reference to the First Principles of Physiology, providing an origin for physiologic evolution (Torday and Rehan, 2009), mimicking the Singularity that preexisted the Big Bang. Over the course of evolution, factors in the environment that posed an existential threat were assimilated by the organism and made useful as our physiology through cell-cell signaling, soluble growth factors being produced by one cell, binding to its receptor on a different cell type.

As such, life may be seen in its entirety, described as the Implicate Order by David Bohm. It is hypothesized that the Implicate Order is what we think of as our unconscious and that our consciousness has emerged from it through our experiential interactions with the environment, giving rise to our subjective senses (Bohm, 1980). Jung referred to this as the "collective unconscious," whereas Gebser characterized it as "the archaic structure of consciousness."

The unicell was the first niche construction (Torday, 2016), assimilating factors in the environment, made useful as physiology. The reference for such factors is the First Principles of Physiology (Torday and Rehan, 2009). When this process is combined with the principle of Terminal Addition—attaching newly acquired traits onto the ends of serial adaptations—it forms a comprehensive means by which the organism orients itself to the Cosmos in what we think of as consciousness. The Cosmos then acts as a lodestone for the organism. It is this set of principles that constitutes consciousness.

How does this interactive relationship between the subconscious Implicate Order and the conscious Explicate inform us? For one thing, it explains Near Death Experiences, Out of Body Experiences, Maslow Peak Experiences, meditation, and the Runner's High, all of which represent direct connections between the Implicate and Explicate Orders. This way of understanding our being also explains how and why the poets, writers, sculptors, painters, and composers are able to capture our soul, short-circuiting the thought process by connecting the Implicate subconscious with the Explicate.

The dual properties of our consciousness as Implicate and Explicate emanate from the function of the cell membrane acting both as a binary for determining whether substances are allowed to enter the cell or not, as the Explicate Order, and at the same time behaving like a Mobius strip, the cell "remembering" its molecular origin as lipids immersed in water before they formed micelles, transcending the Explicate Order, becoming more like the Implicate Order. This is the source of creativity or imagination that distinguishes humans from all other species, as the product of selection pressure for our oversized central nervous system, which enables us to contemplate past, present, and future simultaneously, particularly under conditions of existential threat such as Near Death Experiences, Out of Body Experiences, Maslow Peak Experiences, and the Runner's High. All of these conditions are brought on by wave collapses, calcium fluxes being synchronized by neuroendocrine hormones like oxytocin, vasopressin, and endorphins.

Furthermore, the intersection of physics and physiology would explain the interrelationship between the cell and the knot-derived circle. The circle exists within the Implicate Order, whereas the cell exists within the Explicate Order, but both derive from a common origin. So for example, when Husserl tells us that geometry is actually within us in his "The Origin of Geometry," he is correct. That notion reflects all the way back to Plato's concept of archetypal forms within our minds.

And this novel way of thinking about our interrelationship with the Cosmos also changes the way we think of meditation. Yes, it is a process of introspection, but it naturally transcends the corporeal by delving into our physiologic connection with the Cosmos. And because we add new traits onto a series of evolved traits, referred to as "Terminal Addition," there is an order to our physiology that corresponds with the Cosmos, providing a logic to our interrelationship with it (Torday and Miller Jr., 2018). And the practice of meditation, like every other process, is a learning experience because as Donald Hebb has said, "neurons that fire together, wire together."

Now, when the above way of understanding what consciousness constitutes is combined with the cell as a niche construction, we are able to experience collective consciousness. It is certainly why we believe in something greater than ourselves, reinforced by our institutions, both real and virtual. There are many ways in which we undermine such cohesive forces due to the false narratives we have contrived to cope with our ambiguous origin as negative entropy, which defies the Laws of Nature, and may be the basis for "original sin," but we now have ample knowledge of our true beginnings and means of informing ourselves to correct such misconceptions.

Stafford Beer was of the opinion that this could be accomplished by connecting the "firm" to a pond. He was on the right track, but did not know how to interface the two. Similarly, how Quantum Mechanics interfaces with biology has remained unresolved, referred to as the "explanatory gap." It is acknowledged that Quantum Mechanics is the universal basis of everything in the Cosmos, including life, but how the very small interfaces with the very large, and everything in between, has remained an enigma. However, having reduced physiology to cell-cell communication, a means of doing so has finally been found based on exaptations or preadaptations as the basis for emergence. Using that approach enabled the reduction of complex physiology to the unicell, raising the question as to what the precursor for the cell was. Immersion of lipids in water spontaneously gives rise to micelles, or protocells that provided a "space" that was subsequently filled over the course of evolution based on Symbiogenesis, the co-opting of factors in the environment that posed existential threats. That raised the question as to where Symbiogenesis emerged from, the solution to which was Quantum Entanglement entraining particles within the protocell, "energized" by gravity impinging on the curved surface of the protocell based on Einstein's Field Theory. Gravity is the nonlocal reference for the Quantum Entanglement that occurs within the protocell. But that raises the question as to where gravity and Quantum Entanglement emerged from. The answer is the Singularity that pre-dated the Big Bang. The details of this narrative are related in the following chapters.

References

Bohm, D., 1980. Wholeness and the Implicate Order. Routledge, London.

Roux, E., 2014. The concept of function in modern physiology. J. Physiol. 592, 2245–2249.

Torday, J.S., 2016. The cell as the first niche construction. Biology 5, 19.

Torday, J.S., 2020. Consciousness, redux. Med. Hypotheses 140, 109674.

Torday, J.S., 2021. Cellular evolution as the flow of energy. Prog. Biophys. Mol. Biol. 29. S0079-6107(21)00071-7.

Torday, J.S., Miller Jr., W.B., 2018. Terminal addition in a cellular world. Prog. Biophys. Mol. Biol. 135, 1–10.

Torday, J.S., Rehan, V.K., 2009. Lung evolution as a cipher for physiology. Physiol. Genomics 38, 1–6.

CHAPTER 3

The unicell as the crucible for consciousness as Quantum Entanglement

Introduction

We remain perplexed by the same fundamental philosophical questions about the meaning of life that plagued the Ancient Greeks for lack of a correct perspective on our ontology and epistemology at the interface between physics and physiology. The following is an effort to rectify that.

The consensus is that Quantum Mechanics (QM) is the basis for all things Cosmos. But how and why QM underpins the structure, function, and behavior of lifeforms has been challenging. This concept is usually dismissed based on decoherence, the unlikelihood that the properties that govern QM are compatible with the Newtonian mechanics of everyday life (Zeh, 1970). But that was up until the mechanism of cell-cell communication was invoked, ubiquitously governing ontogeny, phylogeny, and physiology. In that light, dismissing the role of QM would be equivalent to denying that everything in the Cosmos is composed of atoms. Another such fundamental denial is disregarding the evidence for our origins from a Singularity.

It has been hypothesized (see Fig. 1) that the lipid micelle origin of life generated the "dichotomous" interface between David Bohm's Explicate and Implicate Orders, inferring that prior to that, there was only the Implicate Order. As such, the micelle semipermeable membrane formed the basis for all life forms by mediating Symbiogenesis—the appropriation of existential threats in the environment to form what we think of as physiology. As such, the unicell is the "crucible" for determining how to incorporate such epigenetically acquired factors, hypothetically based on Quantum

Quantum Mechanics, Cell-Cell Signaling, and Evolution
https://doi.org/10.1016/B978-0-323-91297-6.00015-1

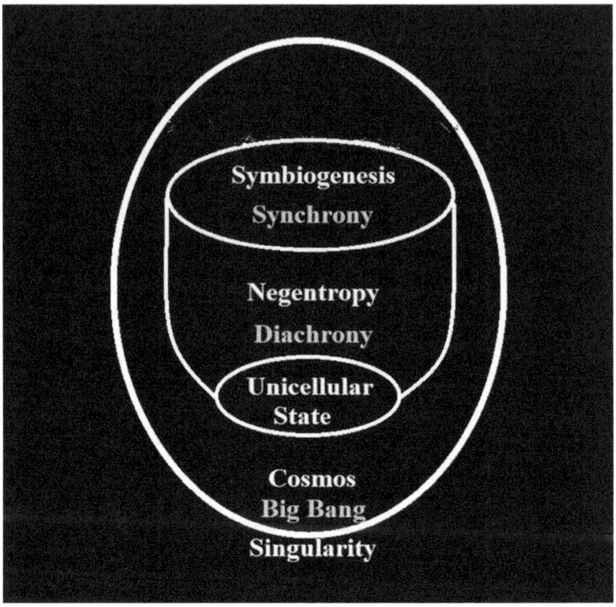

Fig. 1 Schematic for hypothesized micellar origin of life. Beginning with the Singularity (bottom of schematic), moving upward stepwise, the Big Bang gave rise to the Cosmos. The unicell dichotomized the Explicate and Implicate Orders, maintained as negentropy. Quantum Entanglements (see text) were reconciled by nonlocalizations, giving rise to synchronous events as the functional basis for Symbiogenesis.

Entanglements (QEs)—what transpires when particles interact such that the quantum state of each particle cannot be discerned independently of the others—organizing with reference to nonlocalities in the Cosmos. The ultimate cellular determinant for the assimilation of such QE factors is homeostasis, the residual of the "equal and opposite reaction" to the Big Bang, which is Newton's Third Law of Motion. This ontological/epistemological perspective is a frame for the burgeoning field of Quantum Biology.

"Life is that Which Can Mix Oil and Water"—Robert Frost

The cellular basis for the role of QE in evolution is best understood when seen from the origins of the unicellular state. In a series of journal articles and monographs, it has been shown that by reverse-engineering the serially exaptational cell-cell communications for biologic form and function, complicated physiologic traits like lungs, kidneys, and brains can all be traced back to the unicellular state. Using this serial preadaptational or exaptational approach begs the question: Where the unicell itself evolved from? Based on the reverse-engineering approach used, the logical candidate is the Singularity thought to have existed prior to the Big Bang. This perspective allows for a re-telling of the story of evolution from its beginnings instead of its ends, eliminating the systematic error of illogically reasoning after the fact. And it is intriguing to imagine that QE was also the basis, the entire process being founded on QM principles. After all, Rutherford used the Solar System to model his atom, so why not model the planets based on the cell? Applying the schematic in the figure to the Big Bang itself, QM/QE comes to bear at that point in the evolution of everything Cosmos. Occam's Razor would dictate that "one mechanism fits all" is the right answer.

Regarding the process by which the unicell evolved, it has been hypothesized that lipids originating in Pulsars and/or from thermal vents in the sea floor accumulated on the ocean surface. Because they are amphiphiles, the lipid molecules align perpendicularly to the water's surface, their negative poles pointing downward and their positive poles pointing upward. When these lipid molecules pack together, they reduce the surface tension of water caused by Van der Waals forces. Claassen and Spooner have shown that the force of gravity on lipids can then induce the formation of liposomes or micelles. And based on Einstein's Field Theory regarding the effect of gravity on a curved surface such as the micelle, generating energy would have provided the necessary force to maintain negative entropy within the evolving unicell in combination with homeostasis.

As mentioned above, from this point forward, the evolution of unicells over the course of the subsequent 3.5 billion years was due to the endogenization of factors in the environment that posed existential threats. Perhaps initially, before there was a symbiogenic reference, particles entering the protocell assembled based on QE, referencing nonlocal entities as their organizing principles. It was only once the unicell had evolved an organic "self" that it used principles of self-reference to self-organize. Seen in this light, it is no wonder that our physiology is what couples us with the Cosmos.

Further organization of the cells themselves would have been governed by homeostasis, the "equal and opposite reaction" generated by the Big Bang deriving from Newton's Third Law of Motion. It is precisely because the cells formed and were compliant with the same set of Natural Laws as the Cosmos that our physiology runs off of the same operating system, integrating our individual consciousnesses with the Consciousness of the Cosmos. From this perspective, such phenomena as Near-Death Experiences, Out of Body Experiences, and the endorphin "high" can now be explained, instead of being dismissed as qualia.

The vector formed by the Big Bang

The intimate relationship between our personal consciousnesses and the Consciousness of the Cosmos is interconnected through the "vectoral force" produced by the Big Bang. That vector begins with the Elements in the Cosmos, generated by nucleosynthesis of starlight by stars burning deuterium in a serial sequence from the lightest to the heaviest Element, from hydrogen to iron, providing the nonlocal "logic" of the Cosmos; it is because of that Cosmologic logic that the symbiogenic organization of the Elements within the cell subsequently formed a homologous logic, providing the basis for what we conventionally think of as consciousness. This link between ourselves and the Cosmos may constitute the origin of nonlocalization, or what Einstein referred to as "spooky action at a distance." Seen in light of the common origin of consciousness as Cosmology, nonlocalization is no longer a mystical enigma.

Seen synchronically (same space-time), what has been described appears as Darwin's evolution based on adult organisms reproducing based on "fitness," which is teleologic and tautologic. Conversely, seeing the process of evolution as a means for sustaining and perpetuating QEs as the way in which that vector continues to marshal forward is foundational, integrating the inorganic and organic facets of the Big Bang through a common set of Natural Laws.

The above provides the ontology and epistemology for the Cosmos as it pertains to both life and nonlife alike based strictly on empiric evidence. It is predicated on homeostasis not just synchronically as "balance," but diachronically as the "equal and opposite reaction" to the Big Bang. That may also account for the origin of the Laws of Nature, manifested in balanced physical, chemical, and biological reactions alike.

Again, the means by which physiology was deconvoluted all the way back to the unicell was by following the serial cell-cell signaling pathways for homeostatic control. Based on that set of principles, the question arose: "What the exaptation for the unicell was?" If in fact the Big Bang was the universal impetus for everything in the Cosmos, inanimate and animate alike, then the unicell as a "Singularity" may have been patterned after the Singularity thought to have existed prior to the Big Bang. And given the premise that the unicell was the frame for QE, perhaps the Singularity itself was a QE.

Consideration of the Scientific Method bespeaks the validity of the relationship between QE, Symbiogenesis, ontogeny, and phylogeny in the context of evolutionary biology as cell-cell communications. When you conduct a scientific experiment, the criterion for whether the observation is valid or not is contingent on the duplication and replication of the data. That may be congruent with what Carl Jung referred to as "synchronicity," which is the simultaneous occurrence of events that are significantly interrelated. But they do not seem to have any causal connection. However, based on the hypothetical relationship between QE and nonlocalization as the underpinnings of our physiology, perhaps there actually is a causal connection after all. Duplications are virtual synchronicities since they occur simultaneously, and may be Proof of Principle for the interrelationship between QE, nonlocalization, and evolution in the vein of Dobzhansky's statement that "evolution is all of biology" (ontogeny, phylogeny, physiology, etc.). The serial exaptations that have been used to deconvolute evolution are founded on those same experimental duplications and replications.

Observer-observed?

Over the course of the evolution of the brain from the cell membrane, invertebrates had their central nervous systems in their skin, subsequently giving rise to the centralized vertebrate brain. As empiric evidence for this, there are strong molecular homologies between the formation of the lipid barrier produced by the stratum corneum of the skin and the myelination of neurons in the brain. As evidence for that phylogenetic relationship, many neurodegenerative diseases have skin homologues.

Thus, with the "surface" of the Explicate Order behaving as the observer and the Implicate as the observed, the dialectic formed by that interrelationship acts as the "arbiter" for the assimilation of factors in the environment to form physiologic properties. In fact, that is the way that epigenetic marks are incorporated into the egg and sperm, forming methyl adducts of DNA that alter the genetic readout in order to adapt. That differential may ultimately have given rise to lateralization of the cerebral cortex, or left-right brain, acting differentially to identify underlying QM features of the environment. For example, that is why Krishnamurti told Bohm that in order to achieve Cosmic Consciousness, you must relinquish the ego as that membrane-bound structure.

Entanglement forms "knots"

Based on the hypothesis that the unicell determines the intracellular fate of particles acquired from the environment through QEs, over the course of evolution simple cells would have become more and more complicated, resulting in "knots" like those formed in mathematical Knot Theory. The generation of true knots is homologous with the formation of multicellular organisms, both of which comply with the capacity to "unknot" to form the original circle (or cell biologically). This is why the life cycle passes from unicell to unicell, for example. It is, as Husserl states in his classic article "The Origin of Geometry"—"….history is from the start nothing other than the vital movement of the coexistence and the interweaving of original formations and sedimentations of meaning." In the end, knots and cells are surfaces, affording meaning by delineating inside from outside, unlike a Mobius strip as one continuous surface, which has neither.

Cryptochrome, FAD, and magnetoreception

It is now known that certain organisms can geo-locate using "magnetoreception," detecting and following lines of magnetic force across the surface of the Earth. The chemical basis for this trait is cryptochrome, a flavoprotein sensitive to blue light that utilizes FAD (flavin adenine dinucleotide) to produce signals for the brain in order to migrate across great distances. But this only describes the process, but does not determine how and why this particular combination of factors produced magnetoreception. The book "Life on the Edge" (McFadden and Khalili, 2016) by Johnjoe McFadden and Jim Al-Khalili shows numerous synchronic analogies between Newtonian and Quantum Physics, but it does not explain the underlying diachronic homologies (of the same origin) that would lead to an understanding of how and why the unicell formed such interrelationships.

Skepticism about Quantum Biology due to decoherence

As mentioned above, QM properties of life are conventionally dismissed due to decoherence—the inability of QM to interface with the classical physics of physiology. But that fails to recognize the fundamental role of cell-cell communication in ontogeny, phylogeny, and physiology, serving to maintain homeostasis with reference to the Big Bang, i.e., self-referential self-organizational authorship, referencing the Singularity and the QM origin of life. Rejecting the existence of QM in biology due to decoherence is equivalent to rejecting Atomic Theory in physics because it does not connect atoms together.

Discussion

The power to reverse-engineer evolution based on cell-cell communication is highlighted above by its capacity to identify the hypothetical relationship between QM and physiologic principles. Heretofore, it has not been possible to bridge that gap, the argument being that there was no known way to functionally interrelate the two realms. Yet, entertaining the possibility that the micelle provided an intracellular environment in which QE could reconcile nonlocal events accomplishes that "quantum leap." Experimentally, the fact that gravity is necessary for the normal functioning of the cell bespeaks the nonlocal relationship between gravity and life forms. The coupling of gravity with the formation of micelles, and to curvature and energy, provides the a priori underpinnings for the protocell and the First Principles of Physiology (Torday and Rehan, 2009).

It is perhaps too demanding to expect the kind of paradigm shift being proposed herein. But then again, Mankind is faced with huge headwinds such as Climate Change, overpopulation, and food shortages, all of our own making by failing to recognize our ambiguous origins due to negative entropy within the cell, coping by fabricating "Just So Stories." As a result, we are the only species that is destroying the environment, and ourselves in the process. Life began as an ambiguity of entropies within (negative) and outside (positive) of the unicell, and we have been telling ourselves narratives to cope with that condition of uncertainty ever since. And because we are all survivors, those narratives are seen from the perspective of the "winners," feeding our narcissistic tendencies as humans. We must stop deceiving ourselves and begin expressing a narrative consistent with the knowledge we have gained through science if we are to survive as a species.

Even our inventions are a form of deceit, machines providing a mechanical advantage that circumvents the Laws of Nature; it is one thing to assert our power over nature, but we are destroying our ecosystem in doing so at this point. Other organisms manage to live within the boundaries of their abilities, innately knowing that they must constrain their dominance over their environments. We, on the other hand, have evolved our narcissistic behaviors due to the freeing of our forelimbs to make tools, particularly language, oral and written. The Area of Broca is present within the cerebrums of all great apes, from which motor and language skills emanate. We humans are unique among great apes in being bipedal, resulting in excessive selection pressure for the use of our forelimbs to make tools and form language, ultimately giving rise to civilizations that act to protect us so that we may harbor and expand our narcissistic behavior globally, and beyond into outer space. In this vein, it has been postulated that the unicell was the first niche construction as the means of forming a continuum from the environment to the cell; as such, that process would have encompassed the complementary evolution of the individual and civilization as a unified process for the perpetuation and amplification of human narcissism.

The novelty of the present concept for the origin and nature of consciousness derived from the Cosmos is exemplified by Maturana and Varela's book "Autopoiesis and Cognition" (Maturana and Varela, 1980). In it, they state that "The basic cognitive operation we perform as observers is the operation of distinction. By means of this operation we specify a unity as an entity distinct from a background, characterize both unity and background with the properties with which this operation endows them, and specify their separability." This is an eloquent description, but unlike the current explanation for the nature of consciousness based on physiologic evolution, it has remained untestable and irrefutable until now.

This is essentially what has been stated above about the micelle as a crucible, as an "observer" specifying the unity of the Singularity as the origin of the Cosmos. There is a fundamental difference between the two positions in that Maturana and Varela's statement is based on deductive reasoning, whereas what has been proposed herein is based on inductive reasoning, offering the opportunity for experimentation.

Parenthetically, it has been pointed out that basing physiology on function is teleologic and tautologic, yet an alternative has not been forthcoming. The present way of thinking about physiology based on QE and Symbiogenesis offers such a path forward.

Humans are hairless great apes who are born with immature brains in order to allow for the maximum evolution of our oversized heads—over the course of evolution, our heads have grown larger and larger to accommodate our ever-enlarging brains. At some point, the conceptus could not fit through the birth canal any longer. Actually, those forebears who continued along the path of ever-larger heads died in childbirth along with their mothers. Consequently, we survivors are born with only 36% of adult brain volume in order to accommodate brain evolution. As a result, we spend 20–30 years completing the formation of our brains, depending upon civilization to support us. That synergy between our neotenic brains and our narcissistic civilizations has caused massive destruction of our enemies and their environments in the name of humanity. But we are running out of resources in the face of Climate Change, behooving us to modify our behavior accordingly or suffer the consequences.

The ontology and epistemology of life expressed in this chapter offer a novel perspective on the meaning and significance of autopoiesis. We continue on in our material form only as a means for the crucible of QE to reconcile itself with the nonlocal. Armed with such foreknowledge, perhaps we can live life "deliberately" like Henry David Thoreau did when he went to Walden Pond.

References

Maturana, H.R., Varela, F.J., 1980. Autopoiesis and Cognition. D. Reidel, Dordrecht.

McFadden, J., Khalili, J., 2016. Life on the Edge. Crown Publishing, New York.

Torday, J.S., Rehan, V.K., 2009. Lung evolution as a cipher for physiology. Physiol. Genomics 38, 1–6.

Zeh, H.D., 1970. On the interpretation of measurement in quantum theory. Found. Phys. 1, 69–76.

CHAPTER 4

Cybernetics as a conversation with the Cosmos

Prelude

This chapter is framed within the context of Cybernetics as the "feedback mechanism" controlling physiology, from which it originated.

Introduction

Wiener (1948) was the first to define Cybernetics as "the study of control and communication in the animal and the machine." Cybernetics forms the basis for understanding physiology, as described by Cannon (1926), reducing the physical to the biochemical to the cellular-molecular (Boron and Boulpaep, 2021). von Foerster expanded Cybernetics to the Second Order (Glanville, 2021) as awareness of such feedback control, and others have hypothesized Third- and Fourth-Order Cybernetics (Božičnik and Mulej, 2011), but such speculations are not falsifiable, rendering them nonscientific. Here, it is shown that as physiology (Cohn and Cohn, 1953), Cybernetics can be further reduced to cell-cell signaling to maintain homeostasis (Nair et al., 2019), bridging Newtonian Third-Order Cybernetics with Quantum Mechanical Fourth-Order Cybernetics as our "conversation with the Cosmos." That speculation is based on Quantum Entanglement, referencing nonlocalization (Torday, 2021). If everything in the Cosmos, including Quantum Mechanics, is a product of the Big Bang, then the linkage between Third- and Fourth-Order Cybernetics would account for it, and the sooner we begin operating based on Quantum Mechanical principles instead of Newtonian Mechanics, the sooner we will function efficiently based on predictive algorithms, particularly in the realm of Artificial Intelligence.

Quantum Mechanics, Cell-Cell Signaling, and Evolution
https://doi.org/10.1016/B978-0-323-91297-6.00021-7

Cybernetics can be scientifically tested, ranging from the functional to the metaphysical, rendered physically as communication for the first time. This perspective provides common ground for both Science and Art, which are pursued in our never-ending drive to transcend Bohm's Explicate Order, always reaching for the Implicate. From this perspective, consciousness offers a unifying hypothesis for the human endeavor by individuals within social systems to maximize our evolved traits, particularly imagination and abstract creativity, distinguishing us from all other species.

Cybernetics is essentially feedback and is a fundamental way of understanding human physiology (Fedorov, 2007), transcending the present (Torday, 2016a), going stepwise through the history of the organism, and beyond to the very origin of life (Torday, 2019).

It has been theorized that the cell-cell signaling mechanisms that have formed multicellular organisms have occurred through Symbiogenesis (Chapman and Margulis, 1998), Lynn Margulis's theory that when confronted with existential threats over the course of the history of the organism, such factors have been endogenized and assimilated into our physiology (Sagan, 1967), effectively unifying physics and biology. The way in which that occurs is Cybernetic in the context of maintaining or regaining homeostatic control (Torday, 2015), employing soluble growth factors and their receptors as the mediators of cellular awareness of homeostasis (Torday and Rehan, 2007).

But that prompts the question as to what the cell is referencing with regard to homeostasis cybernetically, requiring ingenuity and imagination, given that each step backward over the course of evolution is an exaptation (Gould and Vrba, 1982) or preadaptation (Torday and Rehan, 2007). The question is "What preceded the First Principles of Physiology?" (Torday and Rehan, 2009)—negative entropy, chemiosmosis, and homeostasis. It has been theorized that homeostasis is the "equal and opposite reaction" to the Big Bang, based on Newton's Third Law of Motion (Newton, 2016). Hence, it is the Singularity preceding the Big Bang that is being referenced, given that it is the only entity that existed at that moment in the history of the Cosmos (Hawking, 2020). In that vein, the Cybernetic Q&A "conversation" alluded to in the title of this chapter is between the ever-evolving cell and the Cosmos, or the Fourth Order of Cybernetics.

The origin of life as our Cybernetic "dialog" with the Cosmos

We can trace the Cybernetic principle all the way back from everyday interactions with the environment to Quantum Mechanical principles that connect the unicell with the Cosmos through Quantum Entanglement and nonlocalization. This is how and why David Bohm characterized our existence being between the Implicate and Explicate Orders in "Wholeness and the Implicate Order" as "enfolding and unfolding" (Bohm, 2002), which are homologous with both Symbiogenesis (Margulis, 1993) and niche construction (Torday, 2016b).

With the realization that the divide between self and Cosmos can be seen through the lens of Cybernetics, the lipid-based micelles evolving from lipids immersed in water orienting vertically as amphiphiles with negatively and positively charged ends form the initial boundary between life and nonlife under the auspices of gravity (Claassen and Spooner, 1996). Life is one with the Laws of Nature. Gravity also produces the planets, which distort the fabric of space-time. It has been speculated that the formation of the planets, like that of cells, is also the result of Quantum Entanglement (de Oliveira Neto et al., 2004); thus, planets and cells are homologs, being of the same origin. From that perspective, the origination of life from nonlife due to Symbiogenesis was predictable based on this perspective, in retrospect, neither through deduction nor through induction, but abduction. Life evolves in response to changes in the environment, achieved through cell-environment and cell-cell interactions, mediated by soluble growth factors and their cognate receptors, always referencing the lodestone of the First Principles of Physiology—negentropy, chemiosmosis, and homeostasis—based on experimental evidence (Torday and Miller Jr, 2018; Torday, 2013). Indeed, it is empiricism that has facilitated that continuum, like the Socratic "know thyself" (Benson, 2000), reflected in Maturana's poem...

The Student's Prayer

Don't impose on me what you know,
I want to explore the unknown
And be the source of my own discoveries.
Let the known be my liberation, not my slavery.

The world of your truth can be my limitation;
Your wisdom my negation.
Don't instruct me; let's walk together.
Let my richness begin where yours ends.
Show me so that I can stand
On your shoulders.
Reveal yourself so that I can be
Something different.
You believe that every human being
Can love and create.
I understand, then, your fear
When I ask you to live according to your wisdom.
You will not know who I am
By listening to yourself.
Don't instruct me; let me be.
Your failure is that I be identical to you.

...informing us about the true path of learning as self-awareness through Cybernetic dialog, not "top-down" instruction. This is in the same spirit as Gordon Pask's Conversation Theory and the dialogs between Krishnamurti and Bohm.

If mind is an epiphenomenon of cosmic consciousness, then knowing yourself can be extended into observing the world and acting in the world (mindfully) from the inner dialog between bi-hemispheric aspects of the personal conscious and unconscious. Iain McGilchrist has said "Many 'paradoxes' depend on whether one adopts the left hemisphere (LH) or right hemisphere (RH) way of thinking about, or, more accurately, being in, the world" (McGilchrist, 2016). Extremely different types of attention characterize the LH, which decontextualizes and manipulates, while the RH is complexly interconnected (McGilchrist, 2016); knowing yourself implies knowing your hemispheric limitations (McGilchrist, 2016). To put this in a cellular perspective, some elements (thoughts/emotions) are toxic unless they are integrated into the physiology of an organism.

For example, the iron atom at the core of the hemoglobin molecule facilitates oxygen-carrying capacity, promoting eukaryotic metabolism. In contrast to that, free iron and oxygen are both toxic. However, when each is integrated into the physiology of the organism, it constitutes components of intermediary metabolism. This transition from toxicity to physiology could only occur when such physical substrates are utilized as problem-solving elements symbiogenically. In an information-dependent process in which environmental cues have uncertain meanings due to fluctuating conditions, and information is subject to its own sources of error, the boundary compartment can be seen as a zone in which equivocal information can best be realized. "The default requirements include a self-organized entity that can deploy equivocal information as tools in immediately uncertain conditions based on a set of First Principles that delimits enduring life-sustaining parameters" (Torday and Miller Jr, 2017).

With relation to emotions, "Closely related to emotion regulation are constructs such as mood regulation, coping with stress, and affect regulation. Although it is possible to distinguish semantically between these constructs, their substantive overlap is considerable. At the heart of all emotional states is *core affect* (Russell, 2003), basic states of feeling good or bad, energized or enervated. Regulation of specific emotions, moods, stress, and diffuse affect is therefore always aimed at changing core affect." "Negative emotional states can be costly since they require a wide array of mental and physical resources within the individual" (Sapolsky, 2007). Need-oriented emotion control can thus be adaptive, by allowing individuals to conserve such resources by promoting a rapid return to hedonistically agreeable states (Koole, 2009).

If the purpose of thought and/or emotion is to maintain homeostatic principles, like our cells do, then becoming more mindful of our personal thoughts would provide for a more enduring life-sustaining system. "Homeostasis is constantly oscillating around a set-point, monitoring the cellular environment, always ready to reset itself, but also to provide the reference point for change if necessary for survival in an ever-changing environment." Finding a personal set-point may be a potential goal for divided consciousness, while the personal unconscious is focused on "adaptation of the internal environment… to the external environment of the physical world in service to homeostasis" (Torday, 2015).

Such acts of self-awareness are underpinned by our physiologic "history," commencing with the gyrations of the embryo as it goes from zygote to morula to gastrula, the latter being the stage at which the mesoderm is inserted between the endoderm and ectoderm. Wolpert has said this is the most important thing we will ever do in our lives. And as deuterostomes, we grow and differentiate from anus to mouth, in the opposite direction from the Earth's force of gravity. Moreover, the process of evolving bipedalism recapitulates the vertical orientation of lipids when immersed in water as referred to above, freeing our forelimbs to specialize for flight in the case of birds, and toolmaking and language formation in the case of humans. Fine motor control and language formation are uniquely localized within the Area of Broca in the cerebrums of all great apes, but unlike other members of the species, humans are uniquely bipedal. Bipedalism gave rise to positive selection pressure for an ever-larger central nervous system. Eventually, the conceptus was no longer able to fit through the birth canal, solved by relatively premature birth as hairless apes with only 25% of adult brain volume. Consequently, humans need 20–30 years for our brains to reach their total adult size, accounting for our juvenile behavior—delayed gratification, risk taking, narcissism—which lasts for most of our life cycle; our social and cultural institutions have saved us from ourselves due to our collective conscience.

In the context of such "whipsawing" effects of our evolution, Porges's Polyvagal Theory (Porges, 1995) should also be mentioned. It is formulated based on the development and phylogeny of the vagus, the largest of the autonomic nerves, from the adrenal glands forward to the gut, heart, and ultimately to the cranial nerves that enervate the face. It is this phenomenology that has assimilated the historic "twists and turns" during our evolution that may be the basis for lateralization—left and right brain—acting to accommodate our multifactorial Cybernetic orientation to the Cosmos.

All of the above supports remaining as close to the vector produced by the Big Bang as possible, emanating from the Singularity that preceded it (Wall, 2011). The further we deviate from that vector, the more "friction" there is between us and our evolutionary trajectory, decreasing the possibility of communing with the Cosmos, threatening to break that causal stream of events, which would result in extinction.

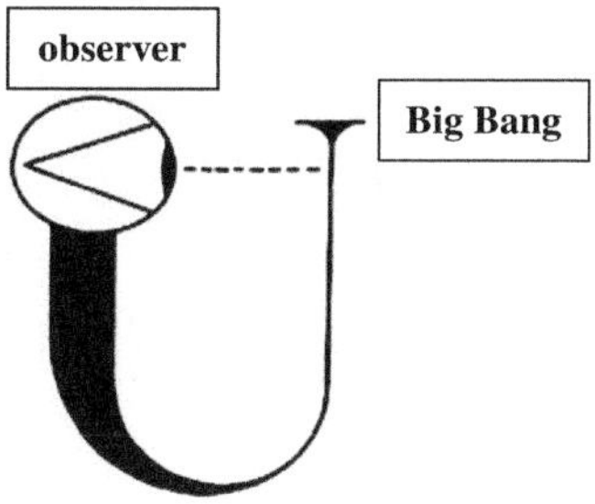

Fig. 1 John Wheeler's "Participatory Universe." Wheeler believed in a universe created by observers.

Big Bang

This perspective is reminiscent of artists and scientists imagining how and why we communicate with the Cosmos, ranging from Robert Frost's "quarrel with the Universe," to John Wheeler's thought that we live in a "Participatory Universe" in which observers create the Universe (see Fig. 1).

However, virtually every way of thinking about our existence tends to enmire us in the synchronic "here and now" view of the world for lack of a way to systematically transcend the present. In order to glimpse the diachronic across space-time perspective, the cell-cell communication mediated by soluble growth factors and their receptors highlighted in this chapter recognizes our "history," offering the possibility of transcending space-time materially, seeing the energetic flow of communications as the true underlying process of life. Mendeleyev was able to accomplish this for chemistry by regressing the synchronic physical characteristics of the Elements against their diachronic atomic numbers, referencing their Big Bang origin (Torday, 2012). Perhaps the same set of principles could be used to devise a Periodic Table for Biology (Torday, 2004).

Physiology as Ross Ashby's Black Box of Cybernetics

Ashby (2015) used Maxwell's Black Box (Glanville, 2008) to explain Cybernetics. It contains a mechanism that cannot be seen and is the product of the observer's interaction with the box. Ashby proposed that the Black Box was not just useful, but is universal, suggesting that we never actually see what's causing a change, only some explanatory principle that think of as a mechanism. This idea of essential obscurity was remarkable at the time and still causes problems for many. Roux (2014) and Cowan (2016) have independently called question attention to the misuse of function to define physiology, for example, arguing that it is teleologic and tautologic to do so. It is no wonder that the definition of physiology remains equivocal given the Black Box nature of Cybernetics as its essence (Iacoviello, 2019).

(Non)conscious or (sub)conscious Cybernetics

We are not always conscious of the Cybernetic relationships beyond the First Order, yet we know that the other three Orders exist based on empiric knowledge of physiology, development, and phylogeny, as well as their recapitulation during injury-repair. (Non)conscious and (un)conscious processes occur in the background of human consciousness. The reason for putting the (non) and (un) in parentheses is to infer a sense that consciousness and unconsciousness continuously intermingle, as life and nonlife are woven together. When Edgar Allan Poe says, "All that we see or seem/Is but a dream within a dream," it congers up a sense of the epiphenomenon of mind produced over generations from prehuman sources.

You are not wrong, who deem
That my days have been a dream;
Yet if hope has flown away
In a night, or in a day,
In a vision, or in none,
Is it therefore the less gone?
All that we see or seem
Is but a dream within a dream.

(Poe, E. A. A dream within a dream. Poetry Foundation. https://www.poetryfoundation.org/poems/52829/a-dream-within-a-dream)

(Non)conscious and (sub)conscious levels of Cybernetics may come to mind during meditation (Liu, 2010), or under conditions of physiologic stress—namely, Near Death Experiences (French, 2005), Out of Body Experiences (Blanke, 2004), Maslow Peak Experiences (Maslow, 2015), or in dreams (Löker, 2001). Such experiences can rise to the level of consciousness, depending upon the circumstances. If the individual is aware of subconscious or unconscious levels of Cybernetics, they are more likely to incorporate such insights into their conscious being. This is largely due to physiology being mediated by cell-surface receptors that can be "taught" to up- or down-regulate. Such practices can be positively reinforced individually through education and positively reinforced socially through institutions and systems.

Gebser has said, "becoming co-visible in and through man it [the diaphainon] attests to the new mutation by dimensioning in consciousness are integrated and made 'meaningful'. And as these spatio-temporal unfoldings or exfoliations open the possibility for the conscious emergence of the presence of origin, this presence diaphanously enters the realm of conscious and visible effectiveness from its unconscious and invisible state" (Gebser, 1985).

On the Cybernetics of auto-engineering

The relationship of evolution to Cybernetics has been alluded to above. The mechanism for such adaptive changes is contingent on the force of homeostasis (Torday, 2015) acting to sustain the equipoise of the structure and function involved. Under physiologic stress, excess blood flow through the microcirculation causes local production of reactive oxygen species (ROS) due to shearing of the vessel walls (Torday, 2013). ROS cause gene mutations and duplications; seen in the context of an evolving trait complying with the First Principles of Physiology, dyshomeostasis results in failure of cell-cell signaling, followed by the cells involved recapitulating their developmental and/or phylogenetic motifs to re-establish homeostatic control by means of soluble growth factors and receptors, recapitulating developmental and phylogenetic motifs, in the spirit of Haeckel's Biogenetic Law, but now with a mechanistic explanation. It also entails Spemann's "Organizer," which is a growth factor-receptor mechanism in hindsight. The underlying principle for homeostatic control is Cybernetic feedback. The newly established set-point is determined by the cellular architecture in what we conventionally refer to descriptively as the process of evolutionary adaptation. Of course, the net result would be Darwinian reproductive success, but that would be an epiphenomenon.

A classic example of this way of understanding evolution at the cellular-molecular level as Cybernetic feedback is recounted in Carroll et al.'s article describing the evolution of the Glucocorticoid Receptor (GR) from the Mineralocorticoid Receptor (Carroll et al., 2011) empirically, due to Cybernetic "trial and error." Yet the authors characterize it in Darwinian terms as a random gene mutation, whereas it occurred during the water-land transition, an extremely stressful experience linking the mineralocorticoid hormone for water and electrolyte regulation, to the hormone for glucose regulation, i.e., glucocorticoids. Additionally, it has been recounted that the physiologic stress of terrestrial life would have increased blood pressure (Kneussl, 1986), and to offset that, the evolution of the GR would have had a dual advantage of decreasing the Mineralocorticoid Receptor levels, at the same time stimulating the number of β-adrenergic receptors (βARs), particularly those in the pulmonary micro-circulation, allowing for independent regulation of the pulmonary and systemic circulations. That modification of the circulatory system was existential for land adaptation. Needless to say, this ad hoc adaptation defies the primary tenet of Darwinian evolution, random mutation (Charlesworth et al., 2017). Conversely, it is consistent with the duplications of both the GR and βAR during the water-land transition, begging the question as to whether these were indeed random events or occur within the homeostatic constraints of any given physiologic trait. Had they been random, they could not recapitulate the process in chronic diseases like emphysema and osteoporosis, for example. The question is furthered by the duplication of the Parathyroid Hormone-related Protein (PTHrP) Receptor during that very same epoch since PTHrP is necessary for the development of lung alveoli (Pinheiro et al., 2012).

This clearly exemplifies Cybernetic feedback between the organism and its environment, the breaching of land resulting in an increase in the effect force of gravity relative to floating in water. Therefore, it is not merely coincidence that the organ of gas exchange for buoyancy would have experienced increased selection pressure, morphing into the lung as the organ of gas exchange for terrestrial life. This interrelationship is underscored by the fact that of the two boney fish phenotypes—physostomous and physoclistous—it is specifically the physostomous type that became our ancestor, possessing a swim bladder connected to its esophagus by the pneumatic duct, which is homologous with the trachea of land-dwelling vertebrates. Furthermore, it has been documented that the top 50 genes that mediate zebrafish swim bladder development are identical with those for the development of the mouse lung (Zheng et al., 2011), providing solid molecular evidence for the homology between the swim bladder and lung. Moreover, FGF10 is necessary for the formation of the smooth muscle that sheathes both the pneumatic duct in fish and the trachea in mice (Korzh et al., 2011), further establishing the homology between the swim bladder of fish and the lungs of mammals.

Discussion

In summary, First Order of Cybernetics is feedback; Second-Order Cybernetics is awareness of the feedback; Third-Order Cybernetics is the cell-cell communications that link cells together to maintain homeostasis, connecting the Newtonian Mechanics with the Quantum Mechanical level; and Fourth-Order Cybernetics is the Quantum Mechanical communication with the Cosmos. Recognition that the ultimate way we commune with the Cosmos is at the Quantum Mechanical scale through cell-cell interactions gains insight into aspects of development and phylogeny untenable otherwise. The fact that soluble growth factors and their cell surface receptors link the cells of multicellular organisms together during the development of the offspring, culminating in integrated physiologic systems, now becomes "rational." By realizing that cell-cell signaling bridges the Quantum Mechanical fundament with the Newtonian Mechanisms we think of as reality, we bridge the so-called explanatory gap. In point of fact, cells can do so diachronically across space-time as the ties that bind us to the Singularity (Torday, 2020), for that is the ultimate reference point for the unicell as zygote, reconciling the asymmetries caused by the Big Bang. That is why we recursively return to the unicellular state over the course of the life cycle, which has merely been dogma up until now. It should be mentioned that physics and chemistry do the same, but their "cycles" occur on a much longer time scale, whereas ours are measured in days in the case of May Flies, to centuries in the case of Giant Red Woods.

Reducing consciousness to Cybernetics allows us to merge the processes of inquiry in Science or the Arts together. We scientifically formulate a hypothesis as a question, with a binary "yes or no" answer that is testable and refutable (Popper's Falsifiability). Artistically, we undertake a similar process, but the methods are much more subjective and indirect, yet the same questioning for the "truth" hangs in the balance. Even in sports, for example, when you hit a tennis ball to your opponent, you are asking the Cosmic question as to whether your opponent knows the "answer" to that shot. The fundament of life is to interrogate the environment to detect novel changes that may or may not affect our survival. Using the example of epigenetic inheritance, we are able to assimilate subtle changes into our offspring via biochemical adducts of DNA in the egg and sperm. Such biochemically modified DNA is subsequently incorporated into the embryo at conception and during the subsequent embryonic development. In that way, such newly acquired knowledge apprises the organism of oncoming environmental change in order to adapt and better survive the ever-changing environment.

The characterization of the Fourth Order of Cybernetics as a "conversation with the Cosmos" may be why we universally feel that there is something greater than ourselves (Buber, 1971). Conventionally, the answer is one of spirituality, or a sense of the group, as in Martin Buber's "I and Thou," but perhaps they derive from that Cybernetic Cosmic dialog. Such a universal property would help us feel like we share a common human bond. Lepskiy (2018) speculates along similar lines about the evolution of Cybernetics, but his work is philosophically based, whereas what is being offered here is based on scientific evidence, allowing it to be falsifiable.

Acknowledgment

Many thanks to Rebecca Damien for her advice regarding metaphysical aspects of this manuscript.

References

Ashby, W.R., 2015. An Introduction to Cybernetics. Mertino Fine Books, Eastford.
Benson, H., 2000. Socratic Wisdom. Oxford University Press, Oxford.
Blanke, O., 2004. Out of body experiences and their neural basis: they are linked to multisensory and cognitive processing in the brain. Br. Med. J. 329, 1414–1415.
Bohm, D., 2002. Wholeness and the Implicate Order. Routledge, London.
Boron, W.F., Boulpaep, E.L., 2021. Concise Medical Physiology. Elsevier, New York.
Božičnik, S., Mulej, M., 2011. A new-4th order cybernetics and sustainable future. Kybernetes 40, 670–684.
Buber, M., 1971. I and Thou. Touchstone, New York.
Cannon, W.B., 1926. Physiological Regulation of Normal States: Some Tentative Postulates Concerning Biological Homeostatics. Editions Medicales, Paris.
Carroll, S.M., Ortlund, E.A., Thornton, J.W., 2011. Mechanisms for the evolution of a derived function in the ancestral glucocorticoid receptor. PLoS Genet. 7, e1002117.
Chapman, M.J., Margulis, L., 1998. Morphogenesis by symbiogenesis. Int. Microbiol. 1, 319–326.
Charlesworth, D., Barton, N.H., Charlesworth, B., 2017. The sources of adaptive variation. Proc. Biol. Sci. 284, 20162864.
Claassen, D.E., Spooner, B.S., 1996. Liposome formation in microgravity. Adv. Space Res. 17, 151–160.
Cohn, S.H., Cohn, S.M., 1953. The role of cybernetics in physiology. Sci. Mon. 76, 85.
Cowan, T., 2016. Human Heart, Cosmic Heart. Chelsea Green Publishing, Chelsea.
de Oliveira Neto, M., Maia, L.A., Carneiro, S., 2004. An alternative theoretical approach to describe planetary systems through a Schrodinger-type diffusion equation. Chaos, Solitons Fractals 21, 21–28.
Fedorov, V.I., 2007. Physiology and cybernjetics: the history of mutual penetration of ideas, modern state and perspectives. Usp. Fiziol. Nauk 38, 72–86.
French, C.C., 2005. Near-death experiences in cardiac arrest survivors. The boundaries of consciousness: neurobiology and neuropathology. Prog. Brain Res. 150, 351–367.
Gebser, J., 1985. The Ever-Present Origin. Ohio University Press, Athens.
Glanville, R., 2008. Second Order Cybernetics. Encyclopedia of Life Support Systems, Paris.
Glanville, R., 2021. Second Order Cybernetics. Systems Science and Cybernetics. Encyclopaedia of Life Support Systems, Paris.
Gould, S.J., Vrba, E.S., 1982. Exaptation- a missing term in the science of form. Paleobiology 8, 4–15.
Hawking, S., 2020. A Brief History of Time. Ishi Press, San Rafael.
Iacoviello, D., 2019. Physiological Cybernetics: Methods and Applications. New Developments on Computational Methods and Imaging in Biomechanics and Biomedical Engineering. Springer, New York.
Kneussl, M., 1986. Regulation of the tracheobronchial and lung vascular system: innervation and receptors. Wien. Klin. Wochenschr. 98, 7013–7025.
Koole, S.L., 2009. The psychology of emotion regulation: an integrative review. Cognit. Emot. 23, 4–41.
Korzh, S., Winata, C.L., Zheng, W., Yang, S., Yin, A., Ingham, P., Korzh, V., Gong, Z., 2011. The interaction of epithelial Ihha and mesenchymal Fgf10 in zebrafish esophageal and swimbladder development. Dev. Biol. 359, 262–276.
Lepskiy, V., 2018. Evolution of cybernetics. Kybernetes 47, 249–261.
Liu, L.H., 2010. Cybernetic unconscious. Crit. Inq. 36, 288–320.
Löker, A., 2001. Cognitive-Behavioral Cybernetics of Symptoms, Dreams, Lateralization: Theory, Interpretation, Therapy. Eruasian Books.

Margulis, L., 1993. Origins of species: acquired genomes and individuality. Biosystems 31, 121–125.
Maslow, A., 2015. Toward a Psychology of Being. Sublime Books, Floyd.
McGilchrist, I., 2016. Selving' and union. J. Conscious. Stud. 23, 196–213.
Nair, A., Chauhan, P., Saha, B., Kubatzky, K.F., 2019. Conceptual evolution of cell signaling. Int. J. Mol. Sci. 20, 3292.
Newton, I., 2016. The Principia. University of California Press, Berkeley.
Pinheiro, P.L., Cardoso, J.C., Power, D.M., Canário, A.V., 2012. Functional characterization and evolution of PTH/PTHrP receptors: insights from the chicken. BMC Evol. Biol. 12, 110.
Porges, S.W., 1995. Orienting in a defensive world: mammalian modifications of our evolutionary heritage. A polyvagal theory. Psychophysiology 32, 301–318.
Roux, E., 2014. The concept of function in modern physiology. J. Physiol. 592, 2245–2249.
Russell, J.A., 2003. Core affect and the psychological construction of emotion. Psychol. Rev. 110, 145–172.
Sagan, L., 1967. On the origin of mitosing cells. J. Theor. Biol. 14, 255–274.
Sapolsky, R.M., 2007. Stress, stress-related disease, and emotional regulation. In: Gross, J.J. (Ed.), Handbook of Emotion Regulation. The Guilford Press, New York, NY, pp. 606–615. 654 Pages.
Torday, J.S., 2004. A periodic table for biology. Scientist 18, 32–33.
Torday, J.S., 2012. Evolutionary Biology, Cell-Cell Communication and Complex Disease. Wiley, Hoboken.
Torday, J.S., 2013. Evolution and cell physiology. 1. Cell signaling is all of biology. Am. J. Phys. Cell Physiol. 305, C682–C689.
Torday, J.S., 2015. Homeostasis as the mechanism of evolution. Biology 4, 573.
Torday, J.S., 2016a. Heterochrony as diachronically modified cell-cell interactions. Biology 5, 4.
Torday, J.S., 2016b. The cell as the first niche construction. Biology 5, 19.
Torday, J.S., 2019. The singularity of nature. Prog. Biophys. Mol. Biol. 142, 23–31.
Torday, J.S., 2020. The Singularity of Nature. Royal Chemical Society, Cambridge.
Torday, J.S., 2021. Life is a mobius strip. Prog. Biophys. Mol. Biol. 167, 41–45.
Torday, J.S., Miller Jr., W.B., 2017. The resolution of ambiguity as the basis for life: a cellular bridge between Western reductionism and eastern holism. Prog. Biophys. Mol. Biol. 131, 288–297.
Torday, J.S., Miller Jr., W.B., 2018. Unitary physiology. Compr. Physiol. 8, 761–771.
Torday, J.S., Rehan, V.K., 2007. The evolutionary continuum from lung development to homeostasis and repair. Am. J. Phys. Lung Cell. Mol. Phys. 292, L608–L611.
Torday, J.S., Rehan, V.K., 2009. Lung evolution as a cipher for physiology. Physiol. Genomics 38, 1–6.
Wall, M., 2011. The Big Bang: What Really Happened at Our Universe's Birth. The History & Future of the Cosmos. Space.com.
Wiener, N., 1948. Cybernetics: Or Control and Communication in the Animal and the Machine. MIT Press, Cambridge.
Zheng, W., Wang, Z., Collins, J.E., Andrews, R.M., Stemple, D., Gong, Z., 2011. Comparative transcriptome analyses indicate molecular homology of zebrafish swimbladder and mammalian lung. PLoS One 6, e24019.

Further reading

Igna, R., Stefan, S., 2015. Comparing cognitive strategies in the process of emotion regulation. J. Evid. Based Psychother. 15, 251–266 (Cluj-Napoca).

CHAPTER 5

Disambiguation, evolution, and the topology of consciousness

Introduction

It has been hypothesized that consciousness is the sum total of our evolutionary history, forged by ontogeny and phylogeny via cell-cell communication. This way of thinking about our awareness of ourselves and our surroundings offers a solution to Chalmer's "hard question" referencing existential threats from our evolutionary past. The following is a recounting of that process.

How and why evolution works has proven refractory to experimentation—how fish have evolved into Man, for example, because Darwinian evolution is based on random mutations, eliminating the possibility for testing such a mechanism, leaving only correlations and associations.

Ideally, there would be some causal process that could be exploited in order to deconvolute evolution, but none has been forthcoming. That is, up until now, having been introduced to Knot Theory, a subheading of Topology, which concerns itself with how geometric objects maintain themselves under continuous deformations. Knot Theory is the study of mathematical knots, originating from literal physical knots. It has occurred to us that since knots tie things together, they are similar in kind to the cell-cell signaling mechanisms that tie physiologic traits together during development, ultimately tying together the homeostatic forces that hold our physiology together.

Evolution is topologic because it began as a "surface," formed by lipid micelles submersed in the ocean that covered the primordial Earth, forming a boundary between the exterior Cosmos and the interiors of micelles. Such protocells comply with the First Principles of Physiology—negative entropy, chemiosmosis, and homeostasis. Negative entropy runs counter to the positive entropy in the

Quantum Mechanics, Cell-Cell Signaling, and Evolution
https://doi.org/10.1016/B978-0-323-91297-6.00005-9

environment, constituting an ambiguous state of being. The processes of development and phylogeny inform the organism of environmental changes, the unicell twisting and turning in its purposeful efforts to conform with its environment in order to act as a phenotypic agent, collecting epigenetic "marks" that keep the organism informed of environmental changes. Such data are then assimilated by the organism's egg or sperm in order to inform the offspring to adapt.

In the case of mathematical knots, they are also generated from cell-like circles that "twist and turn," held together by theoretical "springs," and can be untied by releasing them, reforming as circles. In the case of physiologic "knots," they "untie" themselves in response to physiologic stresses, ultimately during aging or senescence, returning to the unicellular state as a "circle" over the course of the life cycle. So, like the test of a true mathematical knot being the ability to undo itself back to a circle, the "test" of a biologic knot is the ability for it to revert to its unicellular state as a zygote, and so on, and so on, generation upon generation.

The homology between evolution and knots

As stated in the Introduction, knots are topological surfaces, which is how they relate to evolution—lipids originating either from Pulsars or from thermal vents in the sea floor localizing to the air-water interface (Fig. 1), where they formed micelles, prototypical cells that generated a surface between the Explicate and Implicate Orders described by David Bohm in his book "Wholeness and the Implicate Order," forming what Claude Bernard referred to as the "internal milieu." This is the origin of unicellular organisms, which dominated the Earth for the first 3.5 billion years. Due to competition between prokaryotes and eukaryotes, the latter began forming multicellular organisms through cell-cell communications mediated by growth factor-growth factor receptor signaling mechanisms. Such cellular pathways for growth and differentiation during embryologic development subsequently became the homeostatic basis for physiology, as alluded to above. In the process of doing so, the cells involved produce an extracellular matrix that physically stabilizes them, forming true "knots" that literally tie our physiology together as organisms.

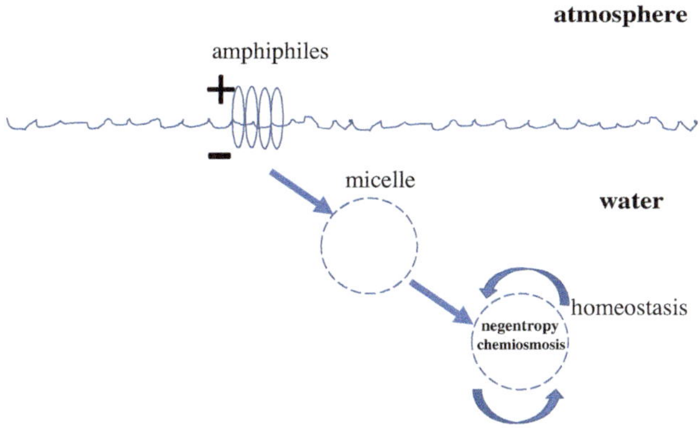

Fig. 1 Formation of micelles from lipids immersed in water. Lipids immersed in water initially align perpendicularly to the surface as amphiphiles, their hydrophobic ends facing upward and their hydrophilic ends facing downward. These lipids will then spontaneously form micelles, spheres with semipermeable membranes surrounding them, because the lipids reduce the otherwise high surface tension of water.

During processes of chronic disease, such cellular communications break down, reverting to earlier developmental and phylogenetic homeostatic stages of evolution, stabilized by scar tissue. During the aging process, loss of bioenergy similarly leads to systematic loss of homeostatic control and deterioration in cell-cell signaling, revealing the knot-like characteristics of organismal structure and function.

Over the course of vertebrate evolution, organisms have coped with their ever-changing environments by endogenizing factors in their environment—heavy metals, gases, ions, bacteria—referred to as the "Endosymbiosis Theory" (Fig. 2). Once such factors were internalized, they were compartmentalized using intracellular membranes, forming what is commonly referred to as "cell physiology." Organisms became progressively more complicated based on cellular communications, producing the phylogenetic changes of speciation. But it should be borne in mind that organisms always return to the

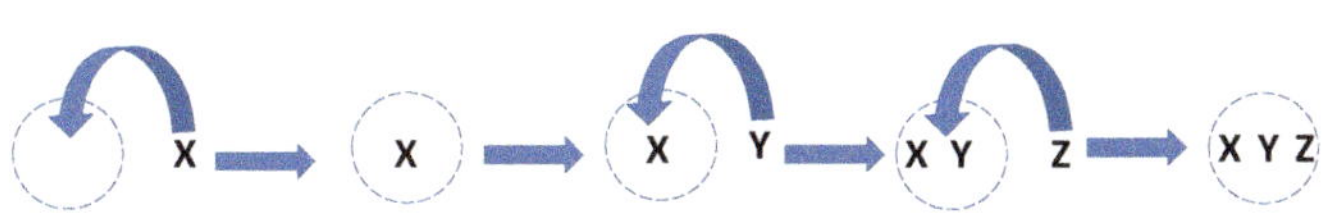

Fig. 2 Endosymbiosis. Over the course of evolution, cells have internalized factors in their environment that have posed existential threats, represented by *x*, *y*, and *z*. The compartmentalization of such factors is the basis for cellular physiology.

unicellular state over the course of their life cycle. The capacity to form and unform structure and function, all the way back to a "circle," is also the criterion for a mathematical knot, attesting to the homology of biological and mathematical knots.

Circles, knots, cells, and memory

Memory is essential to the process of evolution—in order to remain in sync with an ever-changing environment, the organism must remember its past ways of adapting as the most efficient way of surviving, referred to as "exaptations." It is commonly held that DNA is our biologic form of memory, but the use of nucleic acids for this purpose is derivative of the lipid-based micelle, which has molecular memory in the form of hysteresis. The capacity to deform when warmed by the Sun, and reform in the cold of night, is the origin of memory. Like this most basic functional property of life, knots also have memory, because as complicated as they may become, e.g., Trefoils, Borromeans, etc., they remain simple circles at their origin, witnessed by that being the formal validation that they are in fact knots (Fig. 3). The same holds true for a multicellular organism, because it ultimately returns to its unicellular form over the course of its life cycle, demonstrating that it, too, is a knot. By returning to the unicellular state, the organism provides a way of integrating epigenetic "marks" into its physiology over the course of its ontogeny, explaining the cellular basis for evolution.

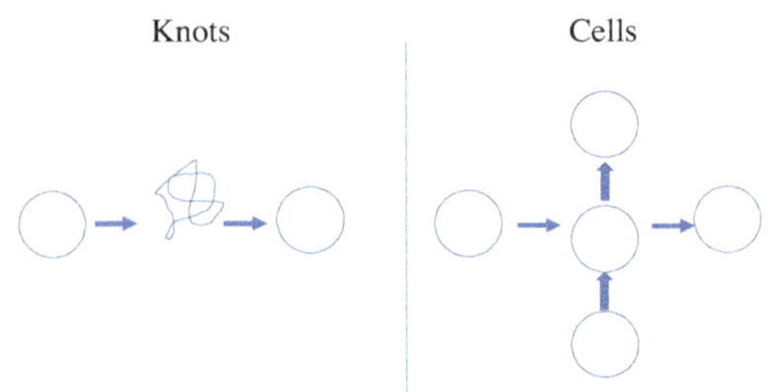

Fig. 3 Relationship between topologic "knots" and cellular evolution. On the left is a schematic for knots as simple circles that can then form complex configurations, but can be unknotted to re-form as circles. On the right is a depiction of cellular evolution, beginning with the unicell (left), forming complicated organisms through vertical integration of cell-cell signaling (middle), ultimately reverting to the unicellular state (right).

And short of regressing all the way back to the unicellular state, tissues can default to earlier stages in their ontogeny and phylogeny due to the loss of cell-cell signaling. This mechanism is due to the cellular networking for development and homeostasis, providing the option of reverting to an earlier physiologic state in order to maintain itself, allowing the organism the opportunity to reproduce and transfer its genome to the next generation.

DNA knots, Homeobox genes, and cell-cell signaling

Evidence that DNA "knots" are altered by changes in cell-cell signaling is not forthcoming. Yet we know that under various stress conditions, cells produce reactive oxygen species that may cause mutations and duplications. Such DNA modifications are then acquired by the genome, hypothetically altering the topology of DNA. The best evidence for this lies in the fact that Homeobox genes, which determine the structure of the organism, are arrayed along the chromosome in the same way that they appear in the organism.

Preadaptations, lipids immersed in water, and knots

The cellular approach to evolution is predicated on serial preadaptations or exaptations, raising the question as to where and when the knots related to life began. In the scenario for the role of lipids in water as the genesis of life, lipids on Earth arose from thermal vents in the sea floor, or as hitch-hikers on the asteroids that delivered frozen water to Earth, the lipids originating from Pulsars, as mentioned above.

Initially, the lipids aligned perpendicularly to the water-air interface (Fig. 1) because they are polarized amphiphiles, their positively charged ends facing upward, being hydrophobic, their negatively charged hydrophilic ends immersed in the water phase. In that configuration, they behave like molecular antennas, responding to electromagnetic forces such as Pulsars and photons of light. When the lipids form micelles, which are spheres with semipermeable membranes, the effect of the electromagnetic force acts to stimulate cytoplasmic streaming as a phenotypic flow of energy. Later on in evolution, when cholesterol appeared in the cell membrane, it caused thinning of the membrane, further facilitating cytoplasmic streaming as locomotion. Fast-forward to *Ciona intestinalis,* a unicellular organism in which the stem cells for the heart develop in the tail and then migrate into the body to form the heart. It is within the realm of possibility to think that the beating of the tail translates into the beating of the heart. This process may have exaptively evolved from the earlier evolution of cytoplasmic streaming since the heart evolved as a muscularization of the vasculature—like the heart of an earthworm, perhaps due to the homologous "cytoplasmic streaming" of blood acting to cause shear stress, known to generate gene mutations and duplications by generating reactive oxygen species (Fig. 4). The one-chambered heart of the earthworm could have given rise to the two-chambered heart of fish, the three-chambered heart of amphibians and reptiles, and ultimately the four-chambered heart of birds and mammals due to β-adrenergic agents like adrenaline. For if you delete the β-adrenergic receptor from developing mice, they are born with a two-chambered heart. The β-adrenergic receptor "duplicated," i.e., amplified during

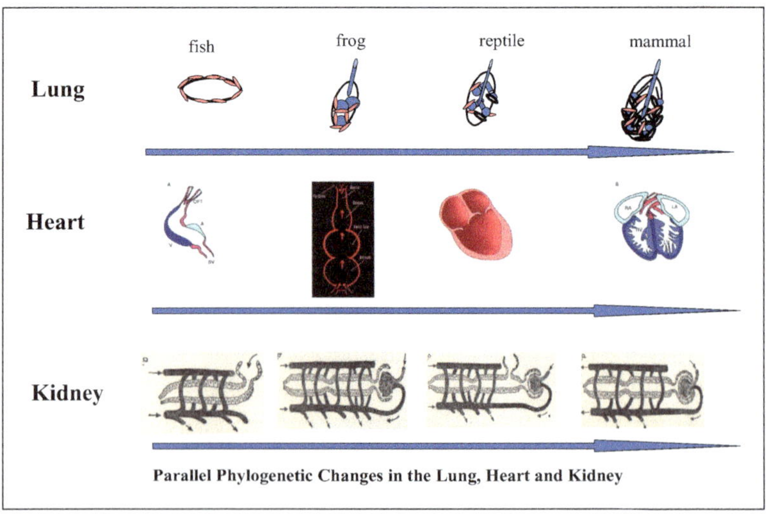

Fig. 4 Evolution of the lung, heart, and kidney. The evolution of the lung (top), heart (middle), and kidney (bottom) is depicted based on developmental and phylogenetic data. Note the increasing complexity of these phenotypic traits from left to right due to the interaction of a changing environment and the cellular components of each structure/function based on cell-cell signaling controlled by homeostasis.

the water-land transition, likely promoting the evolution of the heart from two chambers in fish, to three chambers in amphibians and reptiles, to four chambers in birds and mammals.

The above provides a "knot" that begins with lipids forming a boundary between the Explicate and Implicate Orders, all the way through to the evolution of the lung, heart, and kidney. The genetics of heart development, for example, is connected to those of the fingers, exemplifying the phenotypic homology between these structures based on cytoplasmic streaming.

The deep significance of the homology between knots and evolution

As expressed above, the mechanism of cellular evolution is predicated on exaptations from sequentially earlier and earlier genetic traits, working backward from the present-day physiology to the unicellular state. Starting with lung evolution proved to be advantageous because gas exchange is a very ancient property of life that allowed tracing of the process all the way back to the unicell. But that begs the question, "What the exaptive prototype for the cell itself was?" The only such unity seems to be the putative Singularity of nature that existed prior to the Big Bang. The rationale is that the unicell has the capacity to express all of the states of being—homeostasis, mitosis, and meiosis—all three being determined by the cytoskeleton as the "sensor" for the prevailing environmental conditions. But the insight that knots and evolution share common properties suggests that perhaps even before the homology between the cell and the Singularity, the shared topology of knots and cells may have formed the ultimate homology for the inanimate and the animate.

The metaphysics of knots

The homology of physical and cellular knots provides insight into the continuum from the inanimate to the animate. The phenomenon of lateralization extends that even deeper into our psyche by realizing that our right and left cerebral hemispheres act to perceive the left and right sides of our bodies, respectively, perhaps "unknotting" the

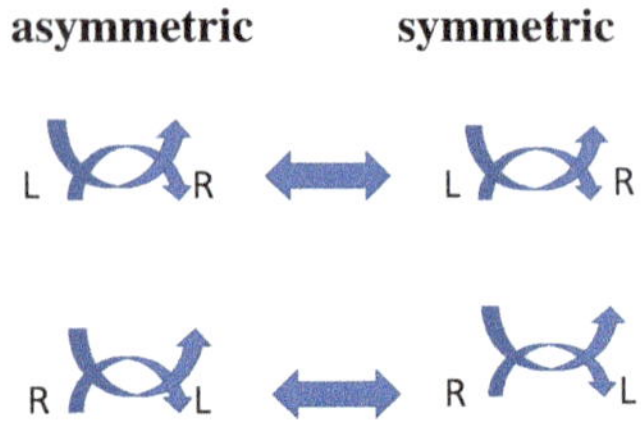

Fig. 5 Lateralization. We perceive things through the asymmetry of our left and right cerebral hemispheres (on the left). However, when we look one another in the eyes, those asymmetries are reconciled (on the right).

ambiguity from which we originated due to the negative entropy that Schrodinger described in "What is Life?" By crossing over from left to right or right to left in perceiving our surroundings, we are able to reconcile that ambiguity and the subjectivity of our senses it gave rise to evolutionarily, seeing the environment in what David Bohm referred to as the "Implicate Order," the true order of things (Fig. 5).

However, when we look one another in the eyes, we transcend that "trompe l'oeil." We instinctively recognize this when a mother looks into her newborn's eyes, or love at first sight, or when we dance together. Perhaps this is what happened to Narcissus when he saw his reflection in a pool of water? Or when an artist creates a novel or a painting or a piece of music? I have experienced insights from time to time while shaving … perhaps it was due to looking in the mirror while holding cold steel to my throat that gave clarity, like what I am expressing here?

Left-right brain as the resolution of Quantum Mechanics in biology: A hypothesis

It has previously been shown that the atom and the cell are homologues, both of which are simultaneously deterministic and probabilistic. In the case of the atom, the nucleus and the electron are in homeostatic balance. Furthermore, the first three quantum numbers for the Pauli Exclusion Principle are deterministic, whereas the fourth principle is probabilistic, based on time. Similarly, the first two First Principles of Physiology—negative entropy and chemiosmosis—are deterministic, whereas the third principle—homeostasis—is probabilistic. Beyond such cellular characteristics, physiology has other

Quantum Mechanical homologies such as nonlocality, coherence, and wave collapse when seen through the "lens" of cell-cell communication as its basis. These insights are particularly significant when it is finally realized that the life cycle is constituted by the unicellular state, transiting from zygote to zygote based on epigenetic inheritance, the organism behaving as an "agent" for the collection of data that modify the DNA readout of the egg or sperm. That perspective offers a much broader understanding of the interrelationship between the animate and the inanimate, leveled by Quantum Mechanics as the universal basis for everything in the Cosmos, bar none. Hypothetically, it is the differential deliberations by the left and right brain based on classic physics that leaves the mathematical "remainder" as Quantum Mechanical. In so doing, such data can then be interpreted by the genome as "meaningful" for inclusion or exclusion to the epigenome.

Discussion

Random gene mutations as the basis for Darwinian evolution disconnect it from development and phylogeny. Conversely, the reverse-engineering of cell-cell communications has offered fundamental insights into the "how and why" of the ontology and epistemology of evolution. The current delineation of the interrelationship between cell-cell communication and Knot Theory further expands on such fundamental causal mechanisms, providing support for the previously expressed concept of "The Singularity of Nature" (Torday, 2019). It is only through such cross-cutting, transcendent diachronies that we will ultimately be able to form an algorithmic perspective for a holistic understanding of life in the Cosmos.

Up until now, it has only been the artists who have been able to express the holism of our existence. For example, Frost said that "life is that which can mix oil and water," or in his poem "The Secret Sits," "… in the middle, and we dance around wondering why." Similarly, in "Little Gidding," T.S. Eliot assures us that-

We shall not cease from exploration
And the end of all our exploring
Will be to arrive where we started
And know the place for the first time.

Fig. 6 Magritte's "Not to be reproduced." *René Magritte. La Reproduction interdite (Not to Be Reproduced). Brussels, 1937.*

In art, Magritte's painting "Not to be Reproduced" (Fig. 6), depicting a man seeing himself in a mirror from behind, expresses the sense of the observer and the observed, while Henry Moore confronts us with the question as to whether the negative space in his sculptures has equal or greater value than the granite or bronze. With the merging of evolution and topology, there is the opportunity to solve these enigmas scientifically and mathematically at long last.

Reference

Torday, J.S., 2019. The singularity of nature. Prog. Biophys. Mol. Biol. 142, 23–31.

CHAPTER 6
On the evolution of evolution

Frescura and Hiley (1984) tell us that when harmonic processes intersect at right angles to one another, they tend to form circularities, but the authors do not explain how or why that should be the case. Ontogeny, or development, and epistemology, or how we gain knowledge, are both harmonic processes that repeat themselves over the history of biology. And ontogeny can be seen as "perpendicular" to phylogeny, each species being formed developmentally over the course of speciation (see Fig. 1).

Haeckel's Biogenetic Law was based on his idea that ontogeny recapitulates phylogeny, but he did not have the scientific data to support this hypothesis, so it fell out of favor. It lay fallow for more than 100 years until it was discovered that the patterns of development are due to cell-cell communications mediated by soluble growth factors and their receptors on target cells, determining the growth and differentiation of embryonic structure and function.

By tracing such pathways of cellular communication backward, in the reverse direction, over the course of time and space, otherwise seemingly complicated physiologic traits terminate in the unicellular state that dominated the Earth for 3.5 billion years. The preadaptation for the cell was the lipids delivered to Earth along with water aboard snowball-like asteroids. When lipids are immersed in water, they align at the air-water interface perpendicularly, their negatively charged ends submerged in the water, their positively charged ends facing upward because these molecules are amphiphiles. When they pack together in this configuration, they break the surface tension of the water caused by Van der Waals force, allowing the force of gravity to reconfigure them as micelles, spheres composed of a lipid membrane that is semipermeable.

Quantum Mechanics, Cell-Cell Signaling, and Evolution
https://doi.org/10.1016/B978-0-323-91297-6.00007-2

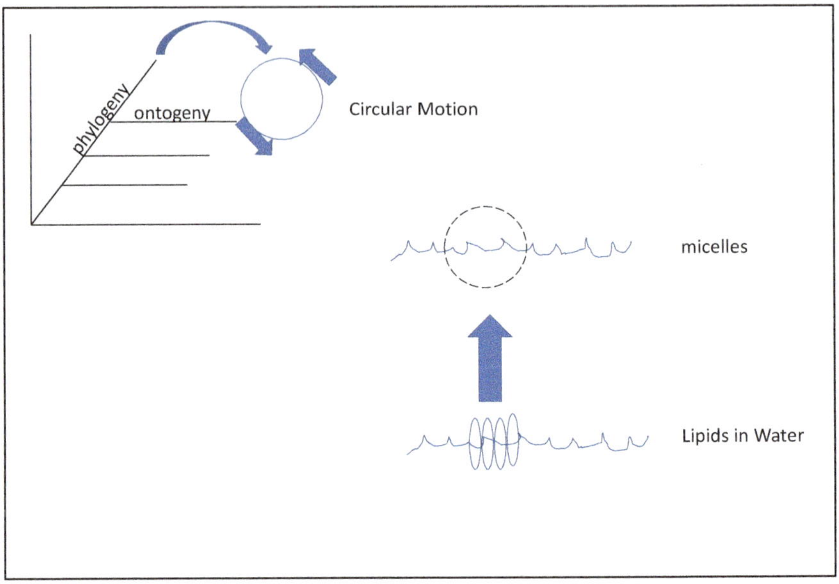

Fig. 1 On the evolution of evolution. Lipids immersed in water form micelles. Micelles were primitive cells formed by the force of gravity, giving rise to ontogeny and phylogeny.

But that still begs the question "What the preadaptation for those micelles was?" They may well have referenced the Singularity that existed before the Big Bang, since it was likely a globular mass. Micelles were the origin of life on Earth, punching a hole in David Bohm's Implicate Order to form the Explicate Order that our subjective senses have formed in order to survive in our Earthly environment. The Explicate Order is therefore a derivative of the Implicate Order. But in terms of life on Earth, it was nonexistent before the Explicate Order was formed. Therefore, everything we experience should be defined in terms of its relevance to the Explicate Order. For example, in his classic article on "The Origin of Geometry," Edmund Husserl intimates that geometric forms are innate to life, not something separate onto itself. Similarly, language and morality are derived from cell-cell communications, not some abstract set of properties.

The concept of self-referential self-organization is central to the above, likely deriving from the amphiphilic nature of lipid molecules, acting to promote such physicochemical behavior in water. One can imagine that when micelles formed, they provided a protected space within which particles could enter or exit due to the semipermeable nature of the lipid membrane. The gravitational force needed to form such micelles would hypothetically have caused Quantum Entanglement of such particles within the cell, whose curved surface would have converted the force of gravity into energy based on Einstein's Field Theory.

Another important aspect of this relationship between lipids and evolution was that they possess molecular "memory." When micelles were warmed by the Sun, they would have liquified, causing them to deform and enlarge their surfaces, allowing for the trapping of more and more calcium from the surrounding water. At night, the micelles would have returned to their original form due to cooling in the absence of the Sun. Eventually, the trapping of calcium would have caused the demise of some micelles due to the denaturing effect of calcium on lipids, but those micelles that devised ways of controlling the flow of calcium became our forebears. That process is the origin of our memory as the flow of calcium, only later being superseded by DNA as the genetic basis for memory.

Subsequent episodes of physiologic stress on biologic systems gave rise to endothermy. The rise in atmospheric carbon dioxide due to plant life caused a "greenhouse" effect, resulting in both a rise in atmospheric temperature and the consequent decrease in the partial pressure of oxygen in water, causing it to be expelled. This combination of effects, referred to as the "Romer Hypothesis," drove boney fish out of water and onto the newly exposed land masses.

Physostomous boney fish have a pneumatic duct connecting their swim bladder to their esophagus, forming the basis for the evolution of the lung. Evidence for this process comes from experiments showing that the swim bladder shares numerous genes with the lung during development. Furthermore, the smooth muscle covering the pneumatic duct and trachea share the same molecular genetics. And both the swim bladder and lung alveoli require surfactant to maintain their physiologic functions.

As fish adapted to land, they experienced specific stresses on their skeleton, skin, and visceral organs, particularly the "need" to breathe air.

During the water-land transition, there were three hormone receptor gene duplications—the Parathyroid Hormone-related Protein Receptor (PTHrPR), the Glucocorticoid Receptor (GR), and the β-Adrenergic Receptor (βAR). All three of these receptors are existential for the effective transition from water to land, and their duplication as receptor genes caused their amplification in response to the stress of adapting to land. For example, the increased relative force of gravity on land versus floating in water likely caused the duplication of the PTHrPR, particularly in combination with the additional stress on the skin to increase barrier function, the lung for breathing, and the kidney for control of fluid and electrolyte balance, all of which were contingent on the amplification of the PTHrPR. In the case of the GR, it evolved from the mineralocorticoid receptor through the effect of shear stress on the microvasculature. In response, the mineralocorticoid receptor was remodeled by adding three amino acids to the binding site for the receptor, now functioning to mediate glucocorticoid signaling as the GR. And in the case of the βAR, selection pressure for independent control of the lung microvasculature was achieved by increasing the number of such receptors due to the emergence of the GR, which upregulates βARs.

Over the course of the transition from the swim bladder to the lung, there would have been intermittent episodes of hypoxia due to the inefficiency of the evolving alveoli. Hypoxia is the most potent stimulus for the hypothalamic-pituitary-adrenal axis, triggering the production of adrenalin by the adrenal gland. Adrenalin would have transiently alleviated the constraint for oxygenation by the alveoli by stimulating the secretion of lung surfactant, allowing the alveoli to further expand due to decreased surface tension. As a "side-effect," adrenalin would also have caused the disruption of fat cells in the periphery, releasing free fatty acids into the circulation. Free fatty acids are the optimal substrate for metabolism, causing increased body temperature. This interrelationship between body temperature and lung surfactant is inscribed in the physiology of evolving vertebrates—the rise in body temperature in mammals is marked by remodeling of lung surfactant to function optimally. The causal relationship between body temperature and lung surfactant was documented by Lau and Keough (1981), who demonstrated that there was a direct effect of ambient water temperature on the composition of lung surfactant in map turtles. And suffice it to say that this interrelationship between lipid metabolism and oxygen economy refers all the way back to those lipid molecules that formed micelles.

Once on land, organisms experienced large fluctuations in atmospheric oxygen, which did not rise from zero to 21%, but in fact rose and fell between 15% and 35%. The increases were marked by the well-recognized gigantisms, whereas the decreases were unappreciated, yet they would have been markedly stressful, perhaps accounting structurally and functionally for the increase in the capillary network of the adrenal medulla. In so doing, it amplified the effect of cortisol produced in the adrenal cortex on the production of adrenalin by the medulla.

The physiologic relevance of this set of evolved traits is underscored by the effect of meditation or hibernation. Such relaxed states result in reductions in blood oxygen, blood pressure, metabolism, and body temperature due predominantly to decreased adrenalin production, decreased release of fatty acids from fat cells, increasing the thickness of the cell membrane, excluding circulating oxygen from the cell, consequently decreasing metabolism and body temperature.

The origination of this process in the orientation of the lipid molecules carries through all of life acting to orient itself to various energetic cues such as gravity, light, and other electromagnetic forces. In plants, these gyrations culminate in gravitropism driving the root into the soil; in humans, it culminates in the lateralization of the brain, forming the left-right brain. In either case, such deep relationships of life with physics reference the underlying Quantum Mechanical basis for life as the common denominator for all things Cosmological.

Seen from this perspective, biology predicts relevant aspects of both life and nonlife. In the reduction of physiology to the unicell, cell-cell communications are necessary for integrated physiology. Loss of such signaling mechanisms results in dissociation of the cell from its phenotype, i.e., identity, indicating that life is determined by its physical relationship to its environment. Furthermore, life forms have evolved by internalizing factors in their environment that have posed existential threats, such as iron, a powerful oxidizing agent, becoming the core of the heme protein within the red blood cell, charged with transporting oxygen within the body. Or oxygen itself, necessary for metabolism of food stuffs, being compartmentalized by lipid membranes throughout the body in order to prevent oxidative destruction of cells. The interrelationship between lipids and oxygen is continuous from the micelle to lung surfactant, acting to constrain oxygen while making use of it physiologically.

References

Frescura, F.A.M., Hiley, B.J., 1984. Algebras, quantum theory and pre-space. Rev. Bras. Fis., 49–86. Vol Especiale.

Lau, M.J., Keough, K.M., 1981. Lipid composition of lung and lung lavage fluid from map turtles (Malaclemys geographica) maintained at different environmental temperatures. Can. J. Biochem. 59, 208–219.

CHAPTER 7

The continuum from physics to cosmology

Introduction

A continuum from physics to biology was first suggested by the Greek Philosopher Heraclitus. Proving the existence of such a continuum has thus far proven untenable, while attempts have been made by the Organic Chemist Cairns-Smith, the Engineer LL Whyte, the Physicist Prigogine, and the Polymath Polanyi. This failure to provide direct evidence for a relationship between physics and biology led both Prigogine and Polanyi to say that biology is too complex to be reducible to physics. Despite all of that, the Astrophysicist Lee Smolin has employed Darwin's theory of evolution to explain such cosmologic phenomena as stellar evolution and Black Holes (Smolin, 1997), showing a mechanistic relationship between physics and biology by inference.

Brian Swimme and Tim Freke have suggested a cosmologic continuum for evolution without invoking the role of physiology. More recently, Torday and Rehan (2012) have shown networking in biology for embryologic development, homeostasis, and physiology based on self-referential self-organization based on cell-cell signaling. That empirically based cell biology approach has proven highly effective, predicting such phenomena as endothermy/homeothermy and the First Principles of Physiology (Torday and Rehan, 2009), reformulating such concepts as homeostasis, phenotype, heterochrony, the cell, and the life cycle based on a unifying mechanism of epigenetics.

Previously, it has been proposed that cognition defines life (Shapiro, 2011; Miller, 2016). Consequently, biological development and evolution depend on the self-referential assessment, deployment, and communication of information (Miller, 2017). All such information-dependent actions are experiential. It has been long argued that

Quantum Mechanics, Cell-Cell Signaling, and Evolution
https://doi.org/10.1016/B978-0-323-91297-6.00018-7

everything in the Universe is experiential. Whitehead hypothesized that even the inanimate must be thought of as experiential. He proposed that reality must be regarded as universal "sense-awareness" (Whitehead, 2006). Seen in these terms, each location is an aspect of every other universal location. Numerous scientists, including the Theoretical Physicist David Bohm, have been highly influenced by this perspective. Whitehead and Bohm were searching for an "underlying wholeness" in which all causation would interrelate vertically within an enveloping realm of superimposed implicates and explicates. This unifying perspective was founded on deep rational consideration before the Singularity of the Big Bang was legitimized by finding cosmic microwave background as its echo, and the recent direct confirmation of gravitational waves.

It is now accepted that universal expansion began as a Singularity, and all that supervenes must emanate within its imprint as shared experientiality as a universal inherency. All life is self-referential, self-organizing, and experiential. Granted that all universal properties must have progenitors from within the Singularity and the Big Bang, biological development and its evolution must extend across quantum space-time as an experiential, relational continuum from that origin. Therefore, life must be seen based on experientiality as universal sense—awareness through which everything has intrinsic connection. Biologically, that experientiality is expressed as self-referential awareness defined as the capacity to maintain cellular homeostasis through the assessment of information and its energetic transfer as cell-cell communication.

With the cell as the basic unit of life, evolutionary development is the continuous process of remaining in "experiential" unity with the Singularity. This is achieved seamlessly as a continuum, with all the quantum properties emanating forward from the Big Bang by consistently, actively internalizing the environment in cellular homeostatic equipoise. Consequently, heredity can be viewed as both physical reproduction and the timeless transfer of fundamental universal sense-awareness as a biological counterpart for Whitehead's continuum, every object connecting to every other universal object.

The Big Bang of the Singularity gave rise to both the physical and biologic realms

The Cosmos is roughly 14 billion years old, resulting from a massive explosion of the "Singularity," giving rise to the material realm. In turn, that realm provided the substrate for biology. Smolin has argued that the physical and biological aspects of the Cosmos are based on self-referential self-organization. Whitehead, de Chardin, and Gurdjieff have proffered a continuum from the physical to the biologic, constituting what we think of as consciousness.

Mechanistically, how could that have come about? A clue can be found within the simplicity of Newton's Third Law of Motion, every action having an equal and opposite reaction. It is proposed that self-referential self-organizational life is best understood as the recoil of the Big Bang, in much the same way as Smolin has attributed Black Holes to forming in response to self-organization—if the reaction is equal and opposite to the "action," it would only stand to reason that the reaction would mirror the original. So both the physical and the biological realms are mini-singularities, authored by the Big Bang.

A feasible pathway for quantum evolutionary development through cell-cell interactions

A cohesive biological narrative can be built on the cellular-molecular principles of morphogenesis since they are the only mechanism known to generate biologic form and function. The big breakthrough in understanding the mechanism of embryologic development occurred in the late 1970s, up until which time it was known that there was a so-called organizer, first described by Spemann, and that there was experimental evidence that there were low molecular weight substances that passed between the embryonic layers that were responsible for organogenesis, as demonstrated by Grobstein in organ culture. But then in 1968, it was discovered that fetal lung development could effectively be accelerated by the relatively simple hormone cortisol. A decade later, Barry Smith subsequently made the breakthrough discovery that cortisol's effect on lung epithelial type II cell surfactant production was indirectly mediated by a soluble factor, which he called "Fibroblast Pneumonocyte Factor (FPF)," produced by alveolar fibroblasts. That led to the seminal insight that pattern formation during morphogenesis was determined by cell-cell interactions, mediated by soluble growth factors that activate their cognate cell surface receptors, triggering intracellular signaling cascades mediated by the so-called second messengers for growth and differentiation.

This mechanistic understanding of embryogenesis has now been applied to evolutionary developmental biology, or EvoDevo. Historically, Smocovitis has documented that cell biology was omitted from evolution theory in the mid-19th Century in favor of the theory that gene mutations were the cause of evolution in her book "Unifying Biology." Not until the 21st Century was lung evolution first proposed to have been caused by epigenetic developmental and phylogenetic changes in cell-cell signaling, initiated by the physiologic stresses caused by the water-land transition. This mechanism for vertebrate evolution may have been overlooked because Darwinian evolutionary theorists have concentrated on random genetic mutations, reluctantly considering robust epigenetic mechanisms. However, heritable epigenetic mechanisms predicated on the effect of physiologic stress causing internal selection for evolutionary developmental and phylogenetic changes have been shown based on experimental evidence. This approach was first used to corroborate the physiologic evolutionary mechanism for the developmental and phylogenetic changes in lung evolution. There were specific receptor gene duplications that occurred during the water-land transition some 500 million years ago, each of which was necessary for land adaptation to amplify their physiologic effects on ontogeny and phylogeny. Experimental manipulation of the pathways regulated by those genes is consistent with their causal relationships in the process of lung evolution.

Mayr's proximate and ultimate causes of evolution

The reduction of embryogenesis to networks of growth factors and their cognate receptors has allowed for the identification of homologies with Quantum Mechanics, which reduces biology to a level that coheres with the Singularity of the Big Bang.

Mayr published a paper in 1952 declaring that there was a fundamental difference between the biologic traits underpinning evolution and the process of evolution itself, the former representing the proximate, and the latter, the ultimate causes of evolution. The biological example he used was that of bird migration. Since then, a great deal has been learned about the reproductive physiology of birds. In particular, there is the effect of ambient light on the production of melatonin by the pineal gland—such data offer an uninterrupted continuum for the seasonal changes in environmental light affecting the reproductive physiology of birds based on cell-cell signaling that can be used to explain how and why birds migrate.

Many other cellular-molecular physiologic properties of vertebrates are now known that provide an understanding of evolution, from the lung, to the kidney, skin, and brain. It is now further known that many of these cell-cell interactions are quantum processes that underpin significant biological properties. For example, the European robin can discern the subtle variations of the angle of the Earth's magnetic field relative to its surface rather than its magnetic orientation. The avian "magnetic compass" is dependent on quantum superposition and the Quantum Entanglement of particles. Further still, models of enzymatic reactions for all organisms cannot explain the thousand-fold increases in speed and amplitude of basic physiological processes without invoking quantum effects via long-range electron and hydrogen tunneling. In fact, it is now evident that life on Earth has been dependent on quantum processes since its beginnings. Quantum coherence and entanglement are the active means for the excitation of the light-harvesting complexes used by photosynthetic bacteria.

This unification of Quantum Mechanics and biological substrates obviates the artificial distinction that Mayr had made between the proximate and ultimate causes of evolution. In so doing, it offers the opportunity to understand the underpinning cellular-molecular mechanisms, like Niels Bohr resolving the duality of light as an artifact of its measurement. The debate over whether evolution is gradual or saltatory exemplifies our difficulty in perceiving this inherent duality. In reality, these processes of evolution are one and the same.

Both depend on latent superpositions in a quantum system in which time is a relative variable.

Comparing apples with apples

By reducing vertebrate evolution to the cellular-molecular level, there is opportunity to mechanistically interface biology with the physical environment. Although there is no direct evidence for the molecular origins of life other than the Miller and Urey experiment (1959), subsequent steps in vertebrate evolution are well documented. Konrad Bloch hypothesized that cholesterol was a "molecular fossil" since it required 11 atoms of oxygen to synthesize one molecule of cholesterol, recounting that stage in evolution when oxygen in the atmosphere was adequate for this to occur. Cholesterol became the structural basis for cell surface receptors as lipid rafts that evolved into cell-cell signaling. The accumulation of CO_2 in the atmosphere during the early phase of unicellular vertebrate evolution led to increased calcium in water due to the formation and erosive effect of carbonic acid on rock. The excess levels of calcium in the ocean water caused endoplasmic reticulum stress, epistatically balanced by the evolution of the peroxisome. The "greenhouse effect" caused by excess carbon dioxide in the atmosphere subsequently caused rising air temperatures, partially drying up the water covering the Earth, driving some vertebrates out of the water onto land. The skeletal changes that occurred during the adaptation to land are well documented (Clack, 2012), but the effect of that transition on the visceral organs has been completely ignored mechanistically as an existential aspect of the process of land adaptation.

The experimental deletion of the Parathyroid Hormone-related Protein (PTHrP) gene underscored the role of this bone calcium regulatory hormone in the evolution of the lung, kidney, skin, and brain, in combination with evidence that the PTHrP Receptor gene duplicated during the water-land transition, effectively amplifying it, offering the opportunity to invoke an evolutionary mechanism. The evolutionary consequences for the adaptation to land can be seen in the physiologic stresses on cell-cell communication in various organs—lung, kidney, skin, bone, brain—allowing for the cell-molecular changes that mediated these tissue-level changes for land adaptation. The direct effects of such environmental factors as oxygen and gravity on morphologic changes fostered the conceptual insights into the physical and biologic environments that constitute evolution. This was the first time evolutionary changes were directly attributed to well-documented sequential geophysical changes in the environment.

When lung or bone cells are exposed to microgravity, expression of the PTHrP gene is inhibited, causing a decline in the PTHrP mRNA content of the cell. Putting the cells back in unit gravity restores PTHrP mRNA expression, demonstrating the causal nature of this maneuver. Such quantum effects of gravity on cell biology are now well recognized, mediated by the Target of Rapamycin gene, which is directly controlled by the cytoskeleton.

Epigenetic inheritance infers the primacy of the unicellular state

Lamarck invoked the concept of epigenetic inheritance in the 18th Century, but was unable to provide scientific evidence for the idea. Only recently has evidence for such direct inheritance of epigenetic "marks" from the environment been shown experimentally. Such epigenetic marks acquired from the environment appear in the germ cells as DNA biochemical adducts, which are passed on to the offspring during the process of reproduction. During meiosis, and then, more particularly, at the level of the unicellular zygote, mechanisms sort out which epigenetic marks are to be retained or discarded. So it is the gametes and zygote that determine the epigenetics of the offspring, not the adults, as would be demanded by Darwinian evolution.

Such interactions between organisms and their environments are self-referential "echoes" of the origins of life. As such, they must proceed through previously enumerated First Principles of Physiology. It may be argued that epigenetics can appropriately be regarded as the biologic "equal and opposite" recoil of the biologic Big Bang. The energetic expansion of the Singularity must have had its opposite according to Newton's Third Law of Motion. Reacting to the universal expansion of the Singularity, biology expresses itself as an entropic endogenization of that expansion via epigenetics, which internalizes aspects of the higher external entropic state of the environment within Schrodinger's negentropic cellular milieu. Therefore, just as the Big Bang gave rise to physics and biology, epigenetics adjusts offspring structure and function as a biologic analog of thermodynamic recoil against the Singularity in order to sustain the essential equipoise of Newton's Third Law of Motion. In this way, the sense-awareness of universal experientiality originally postulated by Whitehead and others can be viewed as a self-organizing force whose obligate direction is toward continuous organismal-environmental complementarity. Epigenetic adjustment and inheritance are that active process. And by this self-same epigenetic means, multicellular eukaryotic reproduction begins with an obligatory reiteration through the unicellular form as its own microcosmic Singularity and proceeds to vast, multitrillion cell elaborations as thermodynamic recoil.

Pauli Exclusion Principle: Homolog for the First Principles of Physiology

Mendeleev successfully formulated a predictive Periodic Table of Elements because he used atomic number as the "common denominator" to normalize the data for the Elements. As Harold Morowitz relates in his book "The Emergence of Everything," when the primordium generated by the Big Bang cooled, the particles, electrons, and photons came together and matter emerged. The electrons interacted with nuclei in certain quantum states to populate their designated orbits. This ordered phenomenon is a function of atomic number, which is derived from Quantum Mechanics. The interaction of electrons with nuclei is characterized by four quantum numbers designated as n, the principal quantum number; ℓ, the azimuthal quantum number; mℓ, the magnetic quantum number; and ms, the spin quantum number. The quantum mechanical solutions are the interaction rules, which yield probabilities for the distribution of the electrons around the nucleus. The Pauli Exclusion Principle states that no two electrons in an atom or molecule can have the same four quantum numbers, three in space and one in time.

The Pauli Exclusion Principle explains the arrangement of electrons and nuclei that results in the atomic number that yields the

Periodic Table of Elements, chemical bonding, and the different states of matter. This property of matter helps to explain how and why the whole is not equal to the sum of its parts, given that the Pauli Exclusion Principle predicts the behavior of two or more electrons rather than any one electron in isolation. Furthermore, it can be argued that the quantum state of the first quantum number determines that of the second and third quanta. However, the fourth quantum number is time-based, which it can be argued confers an implicate of "history," a noetic or "knowing" character is conferred to the elements of the Universe.

It is further proposed that the First Principles of Physiology—negentropy, chemiosmosis, and homeostasis—offer a primal connection between biology and the Singularity. Negentropy and chemiosmosis constrain the living circumstances, and homeostasis can be seen as homologous with Free Will, all being interrelated through the self-referential self-organization that was conferred by the Big Bang.

Nonlocality in physics and biology

The concept of nonlocality has been discussed at length by Bohm and Hiley (1993). They emphasize the novelty of the idea in quantum theory of nonlocality; i.e., that a living system cannot be reduced to parts whose basic properties do not depend on the state of the whole system. They show how this approach implies a new universal type of description, in which the canonical form is always supersystem-system-subsystem, which can be thought of as leading into the radically new notion of the unity of the entire Universe.

Biology complies with the same description. It is not apparent when seen from a synchronic or "real-time" vantage-point, but from a diachronic perspective, transcending space and time as a quantum mechanical continuum, it can be understood in the same terms Bohm and Hiley use for physics. Like Bohm, the Physicist Mae-Wan Ho proposed that the reality within Quantum Mechanics is the opening up of the universal system of the superposition of all possible alternatives (Ho, 1994). In a self-referential, self-organizational context, observation resolves those potentialities into a definite state. But as the Einstein, Podolsky, and Rosen paradox also assumes, there are no preset probabilities within that system. Instead of that, the probabilities are determined within the system, including the biologic. A significant implication resides within this construct. Biologically, the combination of results with the living conditions is not compatible with any basic local interaction model because in order to conform, the signals would have to travel faster than light. Therefore, a necessary conclusion is that quantum physics as nonlocality extends to biology.

That such mechanisms are operative as critical aspects of cell-cell communication is not just conjecture. There is substantial research showing the quantum principles of nonlocality, entanglement, and quantum coherence as active biological mechanisms. They include such fundamental processes as electron transfer during photosynthesis, phototransduction for vision, the radical pair mechanisms in magneto-reception used in animal navigation, and quantum electron tunneling for the sense of smell.

There are mechanisms within a self-referential system in which the collapse of superimposed implicates is instantaneously communicated through an interconnected whole, no matter the distance. An instructive example is muscle contraction; without entanglement as quantum coherence, waving our arms would not be possible. That coordinated action occurs instantaneously over nine orders of magnitude, greatly exceeding any known biochemical, bioelectric, or mechanical connectivity. As Ho (1994) cleverly said, "A quantum world is a radically interconnected, interdependent world where every entity, from elementary particle to galaxy, evolves like an organism, entangled with all there is in nature."

There is no more direct an analogy between a physical system guided by physical laws and quantum rules within our actual living circumstance as holobionts representing complex collaborations between the entire microbiome of each eukaryote and its innate cells. Every multicellular eukaryote, or holobiont, is quite well defined as supersystem-system-subsystem. With that connectedness, the fluid transfer of information as cell-cell communication extends across the entire organism by a variety of quantum means.

From this base, a further means of thinking about biology in cellular-molecular terms is exemplified by pleiotropy. In contrast to the probabilistic way of conventionally thinking about pleiotropy as the random expression of genes throughout the organism to generate more than one distinct phenotypic trait, it is actually a deterministic result of the evolution of complex physiology from the unicellular state. Pleiotropisms come about through recombinations and permutations of cell-cell communication established during meiosis, based on the history of the organism, developmentally and phylogenetically, in service to the future existential needs of the organism. Homologies range from the lung to the kidney, skin, brain, thyroid, and pituitary, exemplifying the evolutionarily mechanistic strategy of pleiotropy. The value of this perspective is exemplified by the resolution, for example, of evolutionary gradualism and punctuated equilibrium in much the same way that Bohr was able to resolve the paradoxical wave-particle duality of light as Complementarity. Seen in this way, biology and physics are both nonlocal, acting at all levels to form and maintain their integrated entirety. Therefore, it can be said that there are identifiable homologies between biology and Quantum Mechanics, as a congruence at every level, from the Singularity across every aspect of the Cosmos.

Pleiotropy is conventionally thought of as the same gene expressed in two or more tissues and organs. If widely enough distributed, when challenged physiologically such a "network" forms the equivalent of an electrochemical field due to the synchronicity of the calcium fluxes. Currently, we think of this in terms of allostasis in a synchronic sense, describing the coordinate interrelationships between the various organs of the body in concert to execute various bodily functions. But when seen as a "field," it opens up to a much more robust way of thinking about the interactions of the organism with its environment, more along the lines of Daniel Fels and his colleagues, considering how cells communicate by unconventional means.

Coherence as Schrodinger wave collapse

The concerted actions of calcium waves referred to above are consistent with the concept of "coherence" in physics. As such, calcium wave collapse as biologic action is a manifestation of coherence. This phenomenon is reflective of experiments with microgravity, showing that in pseudo-zero gravity yeast lose their ability to mediate a calcium flow or bud (i.e., reproduce). Collectively, these lost properties of yeast reveal insight into the essence of life itself since the organism can no longer commune with its environment or reproduce as a result of the collapse of the cytoskeleton when calcium flux is impaired. It is the cytoskeleton that determines the physiologic state of the cell as homeostatic, meiotic, or mitotic. Therefore, the cytoskeleton mediates the communication between the cell and its environment. It can be considered that this flux is the foundation of consciousness. The biochemical connection between the cytoskeleton and its internal environment is mediated by the Target of Rapamycin gene, which controls all of the metabolic properties of the cell, which, in turn, supplies critical support to the cytoskeleton.

Similarly, Hameroff has speculated that anesthetics dissociate us from our environment (render us "zombies") because they bind to the catalytic site of the tubulin gene. He and Penrose have formulated a mechanism for consciousness based on the role of the microtubule system of the brain, integrating its activities, forming a network between and within neurons.

Heisenberg Uncertainty Principle facilitates biologic ambiguity

The beginning of life on Earth was mediated by lipids immersed in the primordial ocean, spontaneously forming micelles as prototypical cells that distinguish the internal from the external environment. Membranes within the cell generate energy via chemiosmosis, allowing for the existence of negative entropy within the cell, in contrast to positive entropy outside of the cell. These two differing states constitute energetic ambiguity, since the self-referential assessment of them is based upon the attachment to biological information space, which is characterized as imprecise information. This forms the beginning of an extensive series of ambiguities that characterize the condition of life, despite its overt material form. Life is dependent on information communicated between the cell and its environment, or between cells.

The sources of information are always compromised, either due to imprecision regarding the sender or receiver of any communication, sender/receiver time delays, or degradation by transit through any intervening medium. It is this fuzziness that fuels evolution, perpetually trying to resolve the problems that are presented to the cell, which is habituated to its ambiguous condition. This is homologous with Heisenberg Uncertainty Principle, disallowing the simultaneous determination of position and momentum of matter. When the Big Bang is the source of both the physical and biologic realms, it should be no surprise that there is shared ambiguity in both those domains, rendering them compatible with one another.

Physiology as recursive fractals

Physiology is fractal, in as much as a faithful adherence to the First Principles of Physiology can be identified as a continuum, beginning with physical energetics, proceeding through unicellular organisms, traceable through complex physiology. Beginning with the spontaneous formation of micelles by immersing lipids in water, and the instantiation of the First Principles of Physiology, each iterative step in evolution references those same First Principles—negentropy and chemiosmosis monitored and governed by homeostasis or homeorhesis. Any significant disturbance of cellular equipoise will always be conditional on cellular changes that would reinstate homeostatic-homeorhetic "balance." Such remodeling processes as the evolution of the alveolus, glomerulus, and skeleton were mediated by cell-cell interactions that would alleviate any imbalances, largely resulting in adaptive changes. The evolution of physiology must be seen through the lens of ecological niche construction. Our evolution proceeds through cell-cell communication mediating the physiologic regulation of genes, and heritable reciprocating signals. At the root of this fractal approach to physiology is an appreciation of the ubiquity of the cell membrane that facilitated oxygenation, metabolism, and locomotion alike from its beginnings as the insertion of cholesterol into the cell membrane.

Biology and its evolution are dependent on a cell membrane that demarcates internal and external energy states. Consider that if the operating principle of life is self-referential self-organization, then the cell membrane might act as the "catalyst" for life itself. The ongoing realization of deep homologies in the physiological systems of widely disparate taxa underscores the fractal nature of physiologic processes. To start, a fractal is a mathematical pattern. It is the math that underlies the dynamics of natural systems and it drives the evolution of phenomena via a basic function that repeats across all levels of time and space, producing self-similarity on all levels. Consequently, the similarity of ontogeny and phylogeny is not selection dependent. Rather, ontogeny and phylogeny are one and the same, as first hypothesized by Haeckel, functioning at different time scales. Upon the inspection of molecular traits, whether ontogenetic (within an individual across time) or phylogenetic (across generations of individuals), there are specific, separable sequences on both time scales. The genes expressed earliest in ontogeny (i.e., immediately following conception) are phylogenetically most ancient. The genes expressed late in development are those that have evolved more recently and have a much narrower phylogenetic distribution. When molecular traits are focused on in reverse, their trajectories end in the unicellular state, suggesting that they have a common origin in the First Principles of Physiology. Organismally, this means that the molecular-level dynamics are self-similar in nature to actions at the cellular level, which scale up to produce both organ- and organ system-level interactions that ultimately culminate in holistic physiology. These fractal interrelationships demonstrate a self-similar reiterating pattern of physiology that assists in the adaptation to the external environment. Since the external environment was formed by the Big Bang, it has been argued that physiology mimics the external Universe by forming its own internal "Universe" for which homeostasis can be considered as a self-referential, self-organizing framework.

It can be put forward that this pattern is shared among all living things. For example, Brad Davidson at the University of Arizona has found that the stem cells for the heart in the tunicate *Ciona intestinalis* are derived from its tail, suggesting that the beating of the tail for locomotion has been exapted for beating of the heart, though unicellular organisms do not require a heart or a circulatory system.

By inference, the heart evolved in support of fundamental biologic traits like respiration, metabolism, and locomotion to support multicellular organisms. If so, the heart is derivative of the convergence of locomotion (as cytoplasmic streaming), metabolism, and respiration. And it can be further argued that all of these traits are derivative of the insertion of cholesterol in the cell membrane.

Exaptations such as the evolution of the middle ear bones in vertebrates from the jaw bones of early fishes have generally provided powerful clues to the ancestry of structures and revealed repeating processes of evolution through innovation from preexisting conditions. Similarly, the brain and gut seem to share a history in response to the demand for central control of the evolving viscera. For example, cells derived from the neural crest of the central nervous system migrate to the developing gut, where they aid in the formation of the gastrointestinal system, including the epithelial barrier, microvasculature, and neuronal enervation.

From the above, it is possible to propose specific linkages between fractal physiology and nutrition. Biology entrained energy via semipermeable membranes, promoting the reduction in entropy that is the metabolic driver for evolution, perpetuating self-organizing homeostasis. For example, the entraining of cholesterol in the plasma membrane facilitated endo- and exocytosis by eukaryotes, as well as aerobic respiration, by thinning out the membrane, making it more permeable for gas exchange.

Another process in this context is chemiosmosis. It could be argued that semipermeable membranes allow for the maintenance of ionic gradients that are fundamental to the energy gradients that energize the force of life. Thus, entropy and chemiosmosis are complementary insofar as they are mutually enacted by semipermeable membranes. As these processes evolved, they had to cope with thermodynamics hierarchically. Cholesterol was subsequently exapted to facilitate the formation of lipid rafts, which form the structural basis for cell-cell signaling, culminating in the synthesis of steroid hormones to form the endocrine system. That interrelationship can now be seen as fractal reiterations in evolution, which, over time, led to vertebrates emerging from water onto land. In this manner, a fractal arc of physiologic evolution can be traced from unicellular to multicellular organisms, from simple to complicated physiology. Similarly, Quantum Mechanics homologously links microlevel atomic principles with such macrolevel phenomena as periodicity, nonlocalization, coherence, and Quantum Entanglement.

The narrative version of vertebrate evolution

Identifying the interface between Quantum Mechanics and biology offers the opportunity to understand evolution as one continuous process, beginning with the Big Bang of the Singularity.

The "recoil" of that explosion as Newton's Third Law of Motion hypothetically caused the equal and opposite reaction in the form of self-organization as a mini-Singularity. That phenomenon offered the opportunity for the formation of the first cell from the lipids present in the primordial ocean since lipids spontaneously form micelles when they are immersed in water, or semipermeable membrane-bound spheres. In effect, the reaction to expansion as increased entropy is internalized within the cell as the critical negentropy that permits self-organization as its opposed reaction. Importantly, a part of that counterbalancing internalization includes the embodiment of the universal sense-awareness that Whitehead (2006) proposed. To support the negentropic state, internal or endomembranes partitioned ions on either side of themselves in the process. Peter Mitchell coined the term chemiosmosis, providing the bioenergy needed for negative entropy, the overall process being governed by homeostasis-homeorhesis. Internalization of the environment by the protocell was tantamount to the first niche construction, the "personalization" of the immediate environment by the organism, furthered by endosymbiotic transfers to enable complex eukaryotic life forms, such as ourselves. Thus, it is averred that the iterative endogenization of the environment in reaction to its ever-changing nature constitutes the process of evolution. This is exemplified by epigenetic inheritance, constituting the direct assimilation of epigenetic marks by the egg and sperm during meiosis. The subsequent transfer of the retained epigenetic marks to the zygote, and their adjudication during that critical phase, facilitates their ultimate passage to the offspring.

Consciousness as the aggregate of the inanimate to the animate

The case can be made for the interrelationship between the physical and biologic realms by which both are connected, each having its own "logic." The prospect that self-reference is the interface between the two forms the conduit for the flow of information between the inanimate and the animate. It is referred to in the literature as the "hard" problem, which becomes the very nature of consciousness.

There is an important aspect to any argument that biology must be understood from the Big Bang forward within the dynamical parameters of Newton's Third Law of Motion. If physics is the fundament of biology, then through the constraints of the Third Law of Motion, there must be an inevitable reciprocation from biology to physics. Therefore, it must be considered that biology is able to inform physics.

Conclusions

To read the works of scientists such as Whyte, Morowitz, Capra and Luisi, and Smolin on the one hand, and mystics like de Chardin and Gurdjieff on the other, there is a continuous, ongoing process in nature that would account for all that we see, from rocks to life, from flora to fauna. We are encouraged in this way of thinking by the feasibility of equating mass and energy as $E = mc^2$, or to conceptualize the merging of all of knowledge as "Consilience" (Wilson, 1999).

The genius of life lies within its self-referential self-organization, yielding the first cell that arose as the consilience of physical mechanisms through the First Principles of Physiology. Single-cell division assured that perpetuating dominance. Multicellular organisms devised more complex mechanisms of reproduction, which ensured their own form of organismal-environmental complementarity through robust epigenetic inheritance. The subsequent stages of multicellular life—embryo, fetus, offspring, life cycle—are all dependent upon the epigenetic reinforcement of self-referential self-organization that began with the Big Bang.

Within this integrated cosmologic perspective lies an identifiable oneness between physics and biology. The prior assumption of passive, sterile random mutation and Darwinian selection is thereby contravened. In its place, there is a self-organizing, sense-aware entirety of universal origins. From this empowering stance, all life is one with the Universe, reiterating across quantum space-time through both classical physical interactions and obscure nonlocalities. Apart and together, in quiet proclamation, they have spread out from the Singularity to resonate with Carl Sagan's profound summation, "We are star stuff which has taken its destiny into its own hands."

References

Bohm, D., Hiley, B., 1993. The Undivided Universe: An Ontological Interpretation of Quantum Theory. Routledge, New York.

Clack, J.A., 2012. Gaining Ground. Indiana University Press, Bloomington.

Ho, M.-W., 1994. The Rainbow and the Worm: The Physics of Organisms. World Scientific, Singapore.

Miller, W.B., 2016. Cognition, information fields and hologenomic entanglement: evolution in light and shadow. Biology 5, 21.

Miller Jr., W.B., 2017. Biological information systems: evolution as cognition-based information management. Prog. Biophys. Mol. Biol. 134, 1–26.

Miller, S.L., Urey, H.C., 1959. Organic compound synthesis on the primitive earth. Science 130, 245–251.

Shapiro, J.A., 2011. Evolution. A View from the 21st Century. FT Press, Saddle River.

Smolin, L., 1997. The Life of the Cosmos. Oxford University Press, Oxford.

Torday, J.S., Rehan, V.K., 2009. Lung evolution as a cipher for physiology. Physiol. Genomics 38, 1–6.

Torday, J.S., Rehan, V.K., 2012. Evolutionary Biology, Cell-cell Communication and Complex Disease. Wiley, Hoboken.

Whitehead, A.N., 2006. The Concept of Nature: The Tarner Lectures Delivered in Trinity College. Echo Library, Fairford.

Wilson, E.O., 1999. Consilience. Vantage, New York.

CHAPTER 8
Quantum Mechanics predicts evolution

Introduction

In the literature on "Quantum Biology" in the ongoing effort to apply Quantum Mechanics (QM) to Evolutionary Biology (Sia et al., 2014), nowhere are the shortcomings of descriptive biology more evident. All prior endeavors to apply QM to Evolutionary Biology have entailed a direct "one-to-one" synchronic approach to the overt physiologic properties of biology as the product of evolution, rather than to the latter's ontologic origins and epistemologic causal mechanisms (Torday, 2018). That is analogous with trying to explain the workings of the internal combustion engine of an automobile by applying physics to the wheels or transmission. In contrast to that, the homologous (same origin) application of QM to the origin and causation of evolution at the cellular-molecular level reveals its mechanistic principles and offers the opportunity to authentically understand how and why physics constitutes the basics of biology (Torday and Miller, 2016a).

Reducing biology to QM offers the opportunity for complementary insights into the cellular-molecular level of biology (Torday and Rehan, 2012). For example, it has previously been shown (Fig. 1) that the direction and magnitude of genetic change over the course of lung evolution can be seen as reciprocating internal and external responses to major geochemical stresses in the environment (Torday and Rehan, 2011). Those changes in genetics service the epistatic balancing selection for calcium and lipid homeostasis (Torday and Rehan, 2012). The selection pressure for this mechanism started with the rise in carbon dioxide in the atmosphere due to plant metabolism (Berner, 1990), causing acidification of the ocean waters, forming carbonic acid that leached calcium from the bedrock (Mitchell et al., 2010).

Quantum Mechanics, Cell-Cell Signaling, and Evolution
https://doi.org/10.1016/B978-0-323-91297-6.00012-6

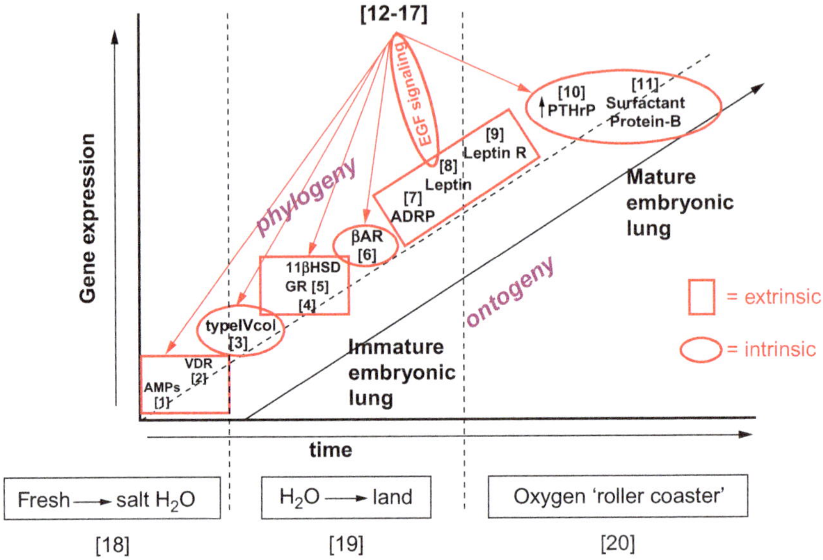

Fig. 1 Reciprocating extrinsic and intrinsic selection pressures for the genes of lung phylogeny and ontogeny. The effects of the extrinsic factors (salinity, land nutrients, and oxygen on the *x*-axis) on genes that determine the phylogeny and ontogeny of the mammalian lung alternate sequentially with the intrinsic genetic factors (*y*-axis), highlighted by the squares and circles, respectively. Steps 1–11 appear in the sequence they appear during phylogeny and ontogeny: (1) AMPs; (2) VDR; (3) type IV collagen; (4) GR; (5) 11βHSD; (6); βAR; (7)ADRP; (8) leptin; (9) leptin receptor; (10) PTHrP; and (11) SP-B. Steps 12–17 represent the pleiotropic effects of leptin on the EGF in oval signaling pathways integrating steps 1–6, 10, and 11. Steps 18–20 are major geologic epochs that have "driven" intrinsic lung evolution. *(Taken from Torday, J.S., Rehan, V.K., 2011. A cell-molecular approach predicts vertebrate evolution. Mol. Biol. Evol. 28, 2973–2981.)*

Calcium can be biologically toxic at high levels, so calcium channels evolved to regulate intracellular calcium (Case et al., 2007). Thereafter, the rise in atmospheric oxygen over the course of the Phanerozoic Era caused endoplasmic reticulum stress, resulting in calcium to leak into the cytoplasm. In reaction, cells formed or acquired peroxisomes, which use lipids to neutralize the deleterious effects of increased cytosolic intracellular calcium (De Duve, 1969). The predictive power of the Quantum Mechanical approach is expressed by the entraining of intracellular calcium, followed by its control through cell-cell interactions mediated by the coordinated effects of soluble growth factors and their receptors, perpetually referencing the First Principles of Physiology (Torday and Rehan, 2012).

Quantum Mechanics and origins and causation in biologic evolution

The consensus is that the Laws of Physics determine all of the Natural Sciences (Birks, 1962), behooving us to answer the question as to how physics has determined the principles of biology. A number of physicists have attempted to solve this riddle (Prigogine and Stengers, 1984; Polanyi, 1968; Schrodinger, 1944) but have failed, Prigogine-Stengers and Polanyi all concluding that biology is just too complicated. Recently, it has been claimed that life is actually simple (Torday, 2016a; Davies and Walker, 2016), but we tend to complicate it with our "Just So Stories" (Barrow and Tipler, 1988), told from our narcissistic, self-serving perspective by reasoning after the fact (Bohm, 1980). Alternatively, there are homologies between physics and biology due to their common origins in the explosion of the Singularity, otherwise referred to as the "Big Bang" (Singh, 2005). Smolin (Smolin, 1997) has effectively argued that stellar evolution and Black Holes came about through Darwinian-like mechanisms. It is possible that subsequent to the Big Bang, based on Newton's Third Law of Motion, an "equal and opposite reaction" was met by self-referential self-organization, given that the only existing paradigm at the time was the Singularity. On that basis, chemistry (Valance, 2017) and biology (Torday and Miller, 2016a) may have come about. There are ways in which we can hypothesize how and why this occurred that are scientifically testable and refutable (Popper, 1963), even though we cannot know the origins, initial conditions, and causal mechanisms of evolution involved because we were not present when they happened. However, it has been argued that since lipids were critically important in the evolution of eukaryotes (Torday and Rehan, 2012), they likely played a vital role in the initial conditions of life based on the premise that evolution is preadaptive, or exaptive, or co-optional (Gould and Vrba, 1982) (Fig. 2). Lipids immersed in water may well have formed the basis for life since they accompanied the frozen snowball asteroids that formed the Earth's oceans (Deamer, 2017), and physicochemically spontaneously form micelles when immersed in water (Hamley, 2007), exhibiting hysteresis as molecular "memory," able to recall their shape and size, which

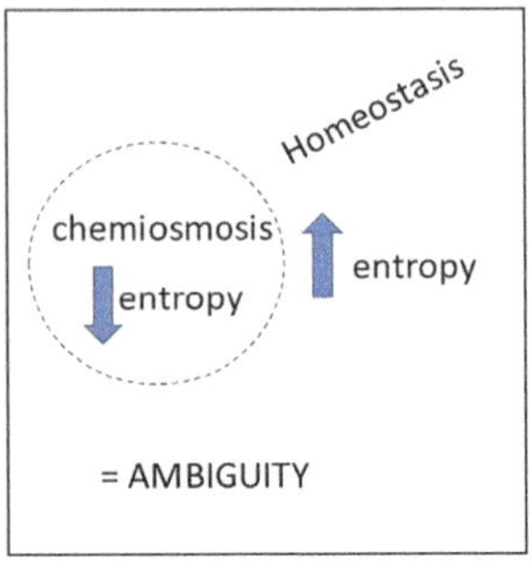

Fig. 2 Origins of life. Lipids immersed in water formed micelles that delimited the internal and external environments, allowing for negative entropy fueled by chemiosmosis within the cell, governed by homeostasis. The differential internal negative and external positive entropies generated the ambiguity of life.

is necessary for evolution (Torday and Miller, 2016). In the process of forming life, the lipid membranes that delimited the internal and external environments generated an ambiguity that became the nidus of life (Torday and Miller, 2017). Under these conditions, several homologs of QM apply, namely, the Pauli Exclusion Principle, nonlocalization, the Heisenberg Uncertainty Principle, and coherence (Krane, 1987). It has been proposed that there are First Principles of Physiology (Torday and Rehan, 2012), namely, negentropy, chemiosmosis, and homeostasis. The first two principles are deterministic, while homeostasis confers Free Will by offering a probabilistic range of potential set points (Bohm and Hiley, 1975). Therefore, life exists within the constraints of determinism and Free Will. Similarly, the Pauli Exclusion Principle generates boundaries of determinism and Free Will in the physical realm since electron spin is ruled by its four quantum numbers, the first determining the second and third when filling the electron's quantum energy state (Krane, 1987). The fourth quantum number is time dependent, providing for a range of probabilistic electron energy levels, conferring Free Will in physical terms. Hence, there is a precedent for determinism and Free Will in the formation of the Elements that biology must comply with. Thus, biology mimics the Quantum Mechanical principles through homologies with the First Principles of Physiology.

Additionally, QM entails nonlocalization, the concept that the Elements of the Cosmos are distributed throughout it (Bohm and

Hiley, 1975). The same property applies to biology in the form of pleiotropy (Torday, 2015). The same gene can be found in many different tissues of the organism as a function of evolution, acting to identify genes in the organism's history that could be used in subsequent novel situations. The recurrence of this process under various environmental conditions constitutes the process of evolution (Torday and Rehan, 2012). Under physiologic stress, the same gene may be expressed in different tissues, always referencing the original unicellular First Principles of Physiology, distributed throughout the organism in a nonlocalized manner (Torday, 2015). This distributive property is homologous with the nonlocalization principle of QM.

Life began as an ambiguity (Torday and Miller, 2017), the entropy of the cell being negative, in contrast to the positive entropy outside of the cell (Schrodinger, 1944). That differential in the internal versus external entropy generated an ambiguity that acts as the ongoing driver for the subsequent evolution of the organism over time both developmentally and phylogenetically. It forms a means of interfacing with the physical environment that exists in a perpetual ambiguous state of uncertainty, allowing the biologic system to effectively cope with the "not knowing with certainty" characteristic of the Heisenberg Uncertainty Principle (Heisenberg, 1949). As an example, Niels Bohr (Baggott, 2011) explained the seemingly paradoxical duality of light as both particle and wave being due to the differences in the way that it is observed, which is homologous with the ambiguity of the cell, coping with such paradoxes, as described above. So perhaps it is because physics, like life itself, exists in a state of limbo regarding its surroundings that allows it to contend with and contemplate it, even at the QM level-like "devolves" like.

The predictive value of a Quantum Mechanical approach to evolution

Prediction is necessary for "hard" science (Popper, 1963). The following is a way of understanding biology mechanistically, allowing it to be interfaced with QM functionally, revealing the common source of both as the Singularity/Big Bang.

The unicell-multicell approach predicts the iterative evolution of niche construction: A holistic approach

The idea that the unicell is a Niche Construction has previously been proposed (Fig. 3) (Torday, 2016b). Physiologic principles have evolved due to the endogenization of the environment, made functional by compartmentalizing (Gabaldon and Pittis, 2015).

Communications between cells supported metabolic cooperativity, to manifesting the principle of Niche Construction in combination with epigenetic inheritance, giving rise to larger and larger organismal communities. The global distribution of these properties culminated in Gaia (Lovelock, 2009), the concept for the organic unity of the Earth. Over the course of the Earth's history, fluctuations in atmospheric oxygen (Berner et al., 2007) and carbon dioxide (Berner, 2003) imposed physiologic stresses on life forms. Systematically tracing the phenotypic changes in the gas exchanger under such regimens revealed the evolution of the mammalian lung from its unicellular origins forward (Torday and Rehan, 2017).

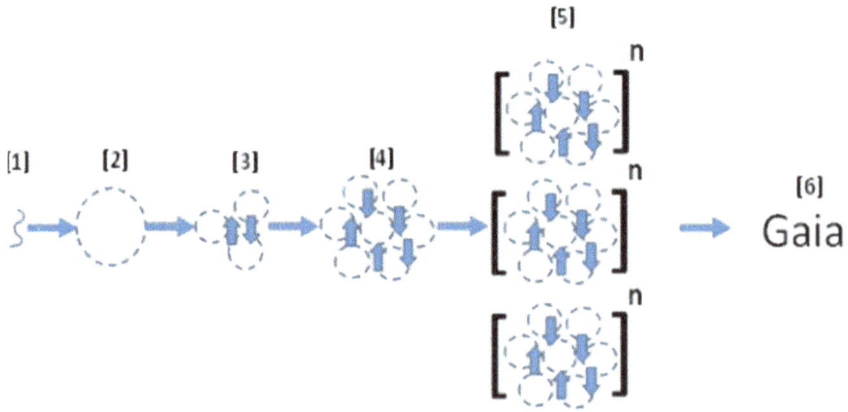

Fig. 3 Niche construction meets epigenetic inheritance. (1) Lipids immersed in water spontaneously form micelles, (2) providing the internal niche construction for the protocell; (3) growth factor signaling mediates cell-cell communication; (4) metabolic cooperativity gives rise to multicellular organisms; (5) interactions between niche construction and epigenetic inheritance foster the formation of organismal communities; (6) global interactions between niche constructions and epigenetics ultimately give rise to Gaia (Torday, 2016b).

The forward-directed continuum of evolution predicts the foundational role of Quantum Mechanics

Viewed from its origins, evolution can literally be seen homologously with QM. Reducing the process of life to its Elements as the First Principles of Physiology (Torday and Rehan, 2012) reveals the one-to-one relationship between QM and physiology, like Mendeleev's Periodic Table, reducing the Elements to their atomic numbers. Therefore, QM can be applied directly to the process of evolution by considering its homologies with the Pauli Exclusion Principle (Pauli, 1925), nonlocalization (Bohm and Hiley, 1975), or Heisenberg Uncertainty Principle (Heisenberg, 1927).

Physiology as a homolog for quantum coherence?

Two waves are coherent when their phase, frequency, and waveform overlap. This is what occurs when either a growth factor, endocrine hormone, or neuroendocrine hormone stimulates its receptor, increasing calcium waves that transduce the signal to affect physiology at the cellular, tissue, organ, or organism level (Schuster et al., 2002).

The cytoskeleton as the homolog for self-organization and consciousness

The cytoskeleton is the superstructure of the cell; actin fibers integrate many aspects of the cytoarchitecture. They anchor proteins and enzymes and keep some membrane proteins in place in specific sites within the plasma membrane such as ion pumps and channels (De Loof, 2016).

The cytoskeleton functions holistically to coordinate all of the functions of the cell in response to the environment as homeostatic, mitotic, or meiotic. In essence, it acts as the homolog for the "consciousness" of the cell. The nature and origins of that property are characterized as the "hard" problem (Harnad, 1998), which may now be understood mechanistically as follows.

Hameroff and Penrose reduce consciousness to the cellular microtubules of the neurons in the brain (Hameroff and Penrose, 2014), acting to transduce its function through the cytoskeleton. This is merely another failed attempt to solve the "hard" problem of consciousness based on descriptive biology. However, seen diachronically back to its origins (Sherrington, 1951), microtubules are Elements of the cytoskeleton, which act as the transducers for all aspects of cell biology—homeostasis, mitogenesis, and meiosis alike (Torday and Rehan, 2012). The structural-functional significance of this interrelationship can be seen in experiments testing the effect of microgravity. For example, when yeast are put in microgravity, they cannot polarize or bud (i.e., reproduce) (Purevdorj-Gage et al., 2006), essentially rendering the organism in suspended animation given that polarization is necessary for calcium flux (Case et al., 2007), and budding is the way yeast reproduce. Similarly, when lung or bone cells are put in microgravity (Torday, 2003), they lose their capacities to express the Parathyroid Hormone-related Protein (PTHrP) gene. PTHrP is a "master" pleiotropic gene determining a wide range of vertebrate traits, from the formation of alveoli (Rubin et al., 1994), to glomeruli (Bosch et al., 1999), to bone calcification (Simmonds and Kovacs, 2010), to skin barrier formation (Philbrick, 1998), and to brain regeneration/function (Gu et al., 2012).

As referenced above, microgravity causes yeast, lung, and bone cells alike to revert to a state of "suspended animation." Hameroff (Hameroff, 1998) has shown that anesthetics have the same net effect physiochemically. Hypothetically, this is the biologic equivalent of QM wave collapse due to the dissociation of consciousness from its physical environment, in this case that of gravity (Schlosshauer, 2005), which is connected to the fabric of space-time (Einstein, 1916).

The central and peripheral mechanisms integrated by PTHrP epitomize holistic physiology, providing insight into consciousness acting to interconnect the cells of the body physiologically. Because of this intimate connection between gravity and physiology, there is a continuum from the inanimate to the animate that was initiated by the Big Bang, to which all of these properties are referenced by QM. Thus, the process of evolution can ultimately be expressed mathematically as a function of QM as it relates to ontogeny and phylogeny.

Discussion

Synchronic versus diachronic

The Periodic Table is an algorithm for understanding the interrelationships between the Elements based on atomic number. The Pauli Exclusion Principle confers both determinism and probability on physics. Like the Periodic Table of physics, a cellular perspective on biologic evolution as the First Principles of Biology provides the mechanistic link to physics, the former similarly indicating that biology exists between the determinism of negentropy and the probability of homeostasis.

The commonality between physics and biology is predicated on self-organization and self-reference as their operating system. Although the origins of these properties are unknown, there is empirical evidence that atoms of yttrium will align themselves spontaneously (Zhang et al., 2017), indicating that there is a mechanism for self-organization.

There is a consensus that the Cosmos resulted from the explosion of the Singularity, known as the Big Bang (Singh, 2005). Newton's Third Law of Motion stipulates that for every action, there is an equal and opposite reaction. Since the only organizational frame of reference known to exist at that time is the Singularity, hypothetically biology is a consequence, acting like a pseudo-Singularity. That would be consistent with what we know of epigenetic inheritance, actively accumulating "marks" that are assimilated by the egg and sperm of the parents, imparted to the offspring during reproduction. This scenario is the reverse of Darwinian evolution, which imbues the adults, not the egg and sperm, with agency for natural selection through mate selection. In contrast to that, like the Red Queen in "Alice in Wonderland," the unicell is doing everything it can to maintain the equipoise possessed by the Singularity—delegating the offspring phenotype to interface with the environment in order to optimally collect data to carry back to the germ cells, zygote, embryo, and offspring of the next generation. Beyond the offspring, the entire life cycle mediates the collection of epigenetic marks since the endocrine system itself is under epigenetic control (Zhang and Ho, 2011). Seeing the process of evolution from the cellular perspective, many dogmas of biology can be understood mechanistically, including heterochrony (Torday, 2016c), homeostasis/homeorhesis (Torday and Rehan, 2012), pleiotropy (Torday, 2015), the life cycle (Torday, 2016a), the cell (Torday, 2016b), and phenotype (Torday and Miller, 2016c).

Like Heliocentrism, this novel perspective on evolution provides a wholly different concept for origins and causation in biology. Conrad Waddington envisioned a horse and cart framed by a window, encouraging us to think of the horse as more than just what we can see here and now. There is an arc of evolutionary history just below the surface within the cellular makeup of the horse (Waddington, 1957). By functionally connecting physics with biology, there is a continuous path from the Singularity/Big Bang to ontogeny and phylogeny as evolution.

References

Baggott, J., 2011. The Quantum Story. Oxford University Press, Oxford.
Barrow, J.D., Tipler, F.J., 1988. The Anthropic Cosmological Principle. Oxford University Press, Oxford.
Berner, R.A., 1990. Atmospheric carbon dioxide levels over phanerozoic time. Science 249, 1382–1386.
Berner, R.A., 2003. The long-term carbon cycle, fossil fuels and atmospheric composition. Nature 426, 323–326.
Berner, R.A., Vandenbrooks, J.M., Ward, P.D., 2007. Evolution. Oxygen and evolution. Science 316, 557–558.
Birks, J.B., 1962. Rutherford at Manchester. Heywood, London.
Bohm, D., 1980. Wholeness and the Implicate Order. Routledge, New York.
Bohm, D., Hiley, B., 1975. On the intuitive understanding of non-locality as implied by quantum theory. Found. Phys. 5, 93–109.
Bosch, R.J., Rodríguez-Puyol, D., Bover, J., Rodríguez-Puyol, M., 1999. Parathyroid hormone-related protein: roles in the glomerulus. Exp. Nephrol. 7, 212–216.
Case, R.M., Eisner, D., Gurney, A., Jones, O., Muallem, S., Verkhratsky, A., 2007. Evolution of calcium homeostasis: from birth of the first cell to an omnipresent signaling system. Cell Calcium 42, 345–350.
Davies, P.C., Walker, S.I., 2016. The hidden simplicity of biology. Rep. Prog. Phys. 79, 102601.
De Duve, C., 1969. Evolution of the peroxisome. Ann. N.Y. Acad. Sci. 168, 369–381.
De Loof, A., 2016. The cell's self-generated "electrome": the biophysical essence of the immaterial dimension of Life? Commun. Integr. Biol. 9, e1197446.
Deamer, D., 2017. The role of lipid membranes in Life's origi. Life 7, 5.
Einstein, A., 1916. Relativity: The Special and General Theory. H. Holt and Company, New York.
Gabaldon, T., Pittis, A.A., 2015. Origin and evolution of metabolic sub-cellular compartmentalization in eukaryotes. Biochimie 119, 262–268.
Gould, S.J., Vrba, E.S., 1982. A missing term in the science of form. Paleobiology 8, 4–15.
Gu, Z., Liu, Y., Zhang, Y., Jin, S., Chen, Q., Goltzman, D., Karaplis, A., Miao, D., 2012. Absence of PTHrP nuclear localization and carboxyl terminus sequences leads to abnormal brain development and function. PLoS One 7, e41542.

Hameroff, S., 1998. Anesthesia, consciousness and hydrophobic pockets—a unitary quantum hypothesis of anesthetic action. Toxicol. Lett., 100–101. 31–39.
Hameroff, S., Penrose, R., 2014. Consciousness in the universe: a review of the 'Orch OR' theory. Phys Life Rev 11, 39–78.
Hamley, I.W., 2007. Introduction to Soft Matter. Wiley, Hoboken.
Harnad, S., 1998. Explaining consciousness: the hard problem. Trends Cogn. Sci. 2, 234–235.
Heisenberg, W., 1927. Über den anschaulichen Inhalt der quantentheoretischen Kinematik und Mechanik. Z. Phys. 43, 172–198.
Heisenberg, W., 1949. The Physical Principles of the Quantum Theory. Dover Books, Mineola.
Krane, K.S., 1987. Introductory Nuclear Physics. Wiley, Hoboken.
Lovelock, J., 2009. The Vanishing Face of Gaia. Basic Books, New York.
Mitchell, M.J., Oliver, E.J., Cliffe, K.A., Maroto-Valer, M., 2010. A model of carbon dioxide dissolution and mineral carbonation kinetics. Proc. R. Soc. A Math. Phys. Eng. Sci. 466, 1265–1290.
Pauli, W., 1925. Über den Zusammenhang des Abschlusses der Elektronengruppen im Atom mit der Komplexstruktur der Spektren. Z. Phys. 31, 765–783.
Philbrick, W.M., 1998. Parathyroid hormone-related protein is a developmental regulatory molecule. Eur. J. Oral Sci. 106, 32–37.
Polanyi, M., 1968. Life's irreducible structure. Live mechanisms and information in DNA are boundary conditions with a sequence of boundaries above them. Science 160, 1308–1312.
Popper, K., 1963. Conjectures and Refutations. Routledge and Kegan Paul, London.
Prigogine, I., Stengers, I., 1984. Order Out of Chaos. Bantam, New York.
Purevdorj-Gage, B., Sheehan, K.B., Hyman, L.E., 2006. Effects of low-shear modeled microgravity on cell function, gene expression, and phenotype in Saccharomyces cerevisiae. Appl. Environ. Microbiol. 72, 4569–4575.
Rubin, L.P., Kifor, O., Hua, J., Brown, E.M., Torday, J.S., 1994. Parathyroid hormone (PTH) and PTH-related protein stimulate surfactant phospholipid synthesis in rat fetal lung, apparently by a mesenchymal-epithelial mechanism. Biochim. Biophys. Acta 1223, 91–100.
Schlosshauer, M., 2005. Decoherence, the measurement problem, and interpretations of quantum mechanics. Rev. Mod. Phys. 76, 1267–1305.
Schrodinger, E., 1944. What Is Life? Cambridge University Press, Cambridge.
Schuster, S., Marhl, M., Höfer, T., 2002. Modelling of simple and complex calcium oscillations. From single-cell responses to intercellular signalling. Eur. J. Biochem. 269, 1333–1355.
Sherrington, C.S., 1951. Man on His Nature. Cambridge University Press, Cambridge. Sia, P.I., Luiten, A.N., Stace, T.M., Wood, J.P., Casson, R.J., 2014. Quantum biology of the retina. Clin Exp Ophthalmol 1957;42:582-589.
Sia, P.I., Luiten, A.N., Stace, T.M., Wood, J.P., Casson, R.J., 2014. Quantum biology of the retina. Clin. Exp. Ophthalmol. 42, 582–589.
Simmonds, C.S., Kovacs, C.S., 2010. Role of parathyroid hormone (PTH) and PTH-related protein (PTHrP) in regulating mineral homeostasis during fetal development. Crit. Rev. Eukaryot. Gene Expr. 20, 235–273.
Singh, S., 2005. Big Bang: The Origins of the Universe. Harper Perennial, New York.
Smolin, L., 1997. The Life of the Cosmos. Oxford University Press, Oxford.
Torday, J.S., 2003. Parathyroid hormone-related protein is a gravisensor in lung and bone cell biology. Adv. Space Res. 32, 1569–1576.
Torday, J.S., 2015. Pleiotropy as the mechanism for evolving novelty: same signal, different result. Biology(Basel) 4, 443–459.
Torday, J.S., 2016a. Life is simple-biologic complexity is an epiphenomenon. Biology(Basel) 5, 17.
Torday, J.S., 2016b. The cell as the first niche construction. Biology(Basel) 5, 19.

Torday, J.S., 2016c. Heterochrony as diachronically modified cell-cell interactions. Biology(Basel) 5, 4.
Torday, J.S., 2018. Quantum mechanics predicts evolutionary biology. Prog. Biophys. Mol. Biol. 135, 11–15.
Torday, J.S., Miller, W.B., 2016a. The unicellular state as a point source in a quantum biological system. Biology 5, 25.
Torday, J.S., Miller, W.B., 2016c. Phenotype as agent for epigenetic inheritance. Biology(Basel) 5, 30.
Torday, J.S., Miller Jr., W.B., 2016. On the evolution of the mammalian brain. Front. Syst. Neurosci. 10, 31.
Torday, J.S., Miller Jr., W.B., 2017. The resolution of ambiguity as the basis for life: a cellular bridge between Western reductionism and Eastern holism. Prog. Biophys. Mol. Biol. 131, 288–297.
Torday, J.S., Rehan, V.K., 2011. A cell-molecular approach predicts vertebrate evolution. Mol. Biol. Evol. 28, 2973–2981.
Torday, J.S., Rehan, V.K., 2012. Evolutionary Biology, Cell-Cell Communication and Complex Disease. Wiley, Hoboken.
Torday, J.S., Rehan, V.K., 2017. Evolution, the Logic of Biology. Wiley-Blackwell, London, London.
Valance, C., 2017. Astrochemistry. World Scientific Publishing Company, Hackensack.
Waddington, C., 1957. The Strategy of the Genes. Routledge, Oxfordshire.
Zhang, X., Ho, S.M., 2011. Epigenetics meets endocrinology. J. Mol. Endocrinol. 46, 11–32.
Zhang, J., Hess, P.W., Kyprianidis, A., Becker, P., Lee, A., Smith, J., Pagano, G., Potirniche, I.-D., Potter, A.C., Vishwanath, A., Yao, N.Y., Monroe, C., 2017. Observation of a discrete time crystal. Nature 543, 217–220.

CHAPTER 9

Goldilocks effect as local paracrine control of homeostasis and endocrinology

Introduction

The endocrine system is an evolved set of hormones that maintain physiologic homeostasis. As such, they monitor the moment-to-moment status of cells, tissues, organs, and whole organism structure and function. Like the process of cellular evolution itself, which senses when there is a loss of homeostasis and remodels itself in order to reinstate cellular-molecular "balance," the endocrine system senses such disturbances and adjusts accordingly. How the various hormonal regulatory mechanisms have evolved is not well known, though there are a few examples that comply with developmental and phylogenetic principles.

Micro- to macroevolution

By reducing the phenotypic algorithms to their smallest functional components and then interrelating them through common mechanisms of morphogenesis, we will be able to drill down to the underlying principles of evolution. Similarly, we could disentangle the evolutionary principles behind the puzzles of lens proteins as crystallins, antifreeze proteins, cytoskeletal proteins, butterfly eyespots, Hox genes, and tetrapod limbs. In contemporary biology and medicine, we are faced with the challenge of merging phenotypic patterns with the genomic and molecular biologic data that undoubtedly generated them. These interrelationships are most apparent during development, because the genetic changes that determine morphogenetic phenotypes are causal. We have suggested that developmental-homeostatic

Quantum Mechanics, Cell-Cell Signaling, and Evolution
https://doi.org/10.1016/B978-0-323-91297-6.00002-3

interrelationships could be exploited to deconvolute the evolution of the lung as a model for the molecular evolution of the physiologic principles described in earlier chapters. The premise that all of biology is a continuum offers the opportunity to exploit that observation in order to determine where it began and how it was perpetuated through evolution. In the present chapter, we have expanded on this concept by integrating the ontogenetic and phylogenetic sequence of genetic mechanisms, regressing them against the temporal sequence of specific geo-environmental milestones in vertebrate evolution: increased ocean salinity, terrestrial nutrients, and atmospheric oxygen. The potential causal relationships delineated in this model have been suggested by their relevance to known developmental mechanisms. The inference is that the ontogeny and phylogeny of the lung have been determined by the impact of extrinsic environmental factors on intrinsic cellular-molecular mechanisms, interacting to generate lung structure and function through positive selection pressure—a process that relates all the way back to the origins of life as micelles forming in response to environmental factors. All of the molecular steps in lung development employed in the model have been shown to be causally related experimentally, and in some cases, the phenomena can be mechanistically explained based on the underlying nature of the mediators. For example, salinity inhibits host defense in fish, but stimulates it through increased vitamin D metabolism, acting as a balancing selection pressure to facilitate evolutionary adaptation. Other examples are the selection pressure for type IV collagen synthesis; the effect of glycyrrhetinic acid, the product of rancidification of land vegetation, on the specialization of the mineralocorticoid and Glucocorticoid Receptors—pentacyclic triterpenoids such as glycyrrhetinic acid inhibiting 11β-hydroxysteroid dehydrogenase (11β HSD1), causing increased blood pressure, resulting in balancing selection for both the Glucocorticoid Receptor and 11β HSD1, resulting in local activation of cortisol within cells and tissues; or the impact of fluctuating oxygen tension in the environment over the past 500 million years on the differentiation of the lipofibroblast to protect the alveolar wall against oxidant injury. It would be expected that if this is positive selection, then the emerging adaptations would also positively interact with the earlier adaptations; in other words, glucocorticoids have a positive effect on the vitamin D receptor (VDR). Oxygen positively affects glucocorticoid signaling and the VDR—perhaps this is seen best as the effect of oxygen on the emergence of the lipofibroblast, producing leptin, which then pleiotropically affects antimicrobial peptides (AMPs), the VDR, type IV collagen, and the surfactant, thereby acting to facilitate the expansion of the blood-gas barrier.

For example, the Glucocorticoid Receptor evolved from the mineralocorticoid receptor as a result of the addition of three specific amino acid residues to the mineralocorticoid hormone binding site (Bridgham et al., 2006). Since this evolutionary change occurred during the water-to-land transition, it has been hypothesized that it was the result of physiologic stress relevant to the adaptive change in question. In the course of the adaptation of boney fish to land, these organisms would have experienced an increase in the effective force of gravity on them on land. As a consequence, there would have been a concomitant increase in blood pressure; in response, the increased shear stress on the blood vessels would have produced reactive oxygen species known to cause genetic mutations. But such mutations would have occurred in the context of cell-cell communications responsible for maintaining homeostasis. In the process of reestablishing homeostasis, the cells would remodel the structures involved and, in so doing, would have modified the mineralocorticoid receptors. Glucocorticoids already existed, but had no specific receptors, but because of their effect on βAdrenergic Receptors, such modifications would have localized to those specific blood vessels experiencing shear stress. Through trial and error, such vessels would have evolved into the Glucocorticoid Receptor through positive selection pressure for the combined effects of the glucocorticoids, offsetting both the stimulatory effect of mineralocorticoids on blood pressure and the positive effect of βAdrenergic Receptors on local regulation of blood pressure. This scenario is consistent with the duplication of the βAdrenergic Receptor during the water-land transition (Aris-Brosou et al., 2009). This mechanism was particularly important in facilitating the water-land transition because it provided the basis for the regulation of the pulmonary blood pressure independent of the systemic blood pressure.

Another such example of hormonal evolution was the modification of the endostyle in elasmobranchs to form the thyroid gland in land vertebrates. The endostyle is an organ in the foregut that traps particulates during feeding. In addition, it also traps bacteria, which stimulate second messengers within the endostyle. Stimulation of cyclic AMP in the endostyle would have caused cellular changes over the course of development and phylogeny consistent with the evolution of the thyroid gland of land vertebrates.

The foregut is a plastic structure from which the thyroid, lung, and pituitary arise through Nkx2.1/TTF-1 expression. Evolutionarily, this is consistent with the concept of *terminal addition*, since the deuterostome gut is formed from the anus to the mouth. Experimentally, when Nkx2.1/TTF-1 is deleted in embryonic mice, the thyroid, lung, and pituitary do not form during embryogenesis, providing experimental evidence for the genetic commonality of all three organs. Their phylogenetic relationship has been traced back to amphioxus, and to cyclostomes, since the larval endostyle, the structural homolog of the thyroid gland, expresses Nkx2.1/TTF-1.

The phylogenetic history of the thyroid

The endostyle is retained in postmetamorphic urochordates and in adult amphioxus. But the postmetamorphic lamprey has a follicular thyroid gland, which is a transformed endostyle. The presence of an endostyle in larval lampreys does not suggest direct descent of lampreys from protochordates, but instead that the evolutionary history of lampreys is long and of ancient origin and that they share the common feature of having filter-feeding mechanisms in their larval stage of development. However, it is notable that the other extant agnathan, the hagfish, possesses thyroid follicles before hatching. Since hagfish evolution is considered to be conservative and hagfish history can be traced back to 550 million years, this suggests that thyroid follicles could likewise be considered to have an ancient history.

Differences in the ontogeny of the thyroid gland is another example of early divergence of lampreys and hagfish during their evolutionary history. However, it is of interest that the method of development of thyroid follicles from a broad pharyngeal epithelium in hagfish embryos is similar to that seen during lamprey metamorphosis, when follicles arise from clumps of cells emanating from the transforming endostyle epithelium. Hagfish development potentially reflects a step in the development of agnathan thyroid follicles that occurred later in lampreys, when metamorphosis appeared in their ontogeny.

The phylogenetic history of the endostyle and nonfollicular thyroid tissues of vertebrates and invertebrates has previously been reviewed by Eales (1997). He emphasized the role of exogenous compounds and peripheral mechanisms in both the regulation and phylogenetic development of the thyroid. Key to thyroid-like function in a tissue is the ability to salvage iodine, and there are many nonchordate invertebrates that have iodine-binding capacity. Ingested and/or absorbed free iodocompounds can be metabolized in the gut assisted by resident bacteria. Absorption of iodocompounds has not been shown in the larval lamprey intestine, but this organ is the primary site for mono-deiodination of thyroxin to triiodothyronine, and this also seems to be the case for ascidians. The endostyle is only seen in marine invertebrates with a notochord and in freshwater larval lampreys. Like many invertebrate tissues, the endostyle has iodine-binding capacity, though this specialized region of the alimentary tract primarily evolved following selection pressure for filter feeding; the iodine-binding capacity of the endostyle in the present-day ascidians and larval lampreys is well documented. Retention of the endostyle throughout lamprey evolutionary history reflects its importance in filter feeding and also suggests that it may have appeared when lampreys were in a pelagic marine phase.

The vertebrate thyroid gland evolved under strong positive selection pressure for the gland during the transition of ancient chordates from the iodine-rich marine environment to the iodine-poor freshwater habitat. In this context, the presence of an endostyle in extant larval lampreys is a reflection of their ancient marine origins, and the adult follicular thyroid gland came after the animal moved toward freshwater. There was positive selection pressure generated by the freshwater environment for metamorphosis in the ontogeny of lampreys, given that the lamprey thyroid only appears following the metamorphic transformation of the endostyle. Transformation of the endostyle into the thyroid gland was a consequence of this metamorphosis.

This view of the ontogeny of the lamprey thyroid gland, along with the belief that the freshwater habitat of lampreys is a secondarily acquired niche, suggests that metamorphosis may not have originated as a developmental process, but instead occurred when lampreys transitioned to freshwater during their evolutionary history. Alternatively, the presence of metamorphosis, and ultimately the follicular thyroid gland, in the marine environment was the primary reason why lampreys could inhabit freshwater. The presence of an endostyle in early lampreys was not necessary for living in a marine environment, since the present-day larvae with endostyles cannot tolerate even dilute seawater. It is only after metamorphosis that juveniles of some species can tolerate seawater.

The larval and reproductive phases of the lamprey life cycle are confined to freshwater. The capacity of juveniles of some species to osmoregulate in seawater may also have been a characteristic that was secondary to the appearance of metamorphosis in the ontogeny of lampreys. An explanation of how marine osmoregulation may have been secondarily derived following the primary derivation of metamorphosis can be found in modern views of the phenomenon termed "developmental integration."

An evolutionary vertical integration of the phylogeny and ontogeny of the thyroid

Mechanistically, the increased bacterial load due to the facilitation of feeding by the endostyle may have stimulated the cyclic AMP-dependent protein kinase A, or PKA, pathway, since bacteria produce endotoxin, a potent PKA agonist. This process may have evolved as the regulation of the thyroid by thyroid-stimulating hormone (TSH), since TSH stimulates the thyroid via the cAMP-dependent PKA signaling pathway. This mechanism hypothetically generated new structures such as the thyroid, lung, and pituitary, all of which are induced by the PKA-dependent Nkx2.1/TTF-1 pathway. Brain-lung-thyroid syndrome, in which infants with Nkx2.1/TTF-1 mutations develop hypotonia, hypothyroidism, and respiratory distress syndrome, or surfactant deficiency disease, is additional evidence for the coevolution of the lung, thyroid, and pituitary.

Developmentally, the thyroid out-pouches from the foregut in the mouse beginning on embryonic day 8.5, one day before the lung and pituitary emerge, suggesting that the thyroid may have been a molecular prototype of the lung during evolution, providing a testable and refutable hypothesis. The thyroid made molecular iodine in the environment bioavailable by binding it to threonine to produce thyroid hormone, whereas the lung made molecular oxygen bioavailable, first by inducing fat cell-like lipofibroblasts as cytoprotectants, which then stimulated surfactant production by producing leptin, placing increased selection pressure on the blood-gas barrier by making the alveoli more compliant. This, in turn, may have created further selection pressure for the metabolic system to utilize the rising oxygen level in the environment, placing further selection pressure on the alveoli, giving rise to the stretch-regulated surfactant system mediated by PTHrP and leptin (Torday and Rehan, 2002). Subsequent selection pressure on the cardiopulmonary system may have facilitated liver evolution, since the progressively increasing size of the heart may have induced precocious liver development, fostering increased glucose regulation. The brain serves as a glucose sink, and there is experimental evidence that increasing glucose during pregnancy increases the size of the developing brain. The further evolution of the brain, specifically the pituitary, would have served to further the evolution of complex physiologic systems.

Both the thyroid and the lung have played similar adaptive roles during vertebrate evolution. The thyroid has facilitated the utility of iodine ingested from the environment, whereas the lung has accommodated the rising oxygen levels during the Phanerozoic Era. In both cases, these structures have accommodated otherwise toxic substances for biologic purposes that have allowed vertebrates to adapt to their environment. Importantly, the thyroid and the lung may have interacted cooperatively in facilitating vertebrate evolution. Thyroid hormone stimulates embryonic lung morphogenesis during development, while also accommodating the increased lipid metabolism needed for surfactant production by driving fatty acids into muscle to increase motility, as opposed to the oxidation of circulating lipids to toxic lipoperoxides. The selection pressure for metabolism was clearly facilitated by the synergy between these foregut derivatives.

Wolff's Law and bone remodeling

Bone tissue can adapt to its environment to optimize its mechanical demands. Wolff's Law was the first to propose that the skeleton can respond to such demands. The cells of the skeleton—the osteoblasts, osteocytes, and osteoclasts—mediate the remodeling process. Studies have focused on how strain affects signaling to regulate bone cells. Regulation of cellular bone remodeling entails both the sensing and response to the mechanical environment. A certain degree of loading is necessary to see the effects of such changes, and the degree of the response is determined not only by the magnitude of the load but also by the rate, the cycle number, and even the frequency of the load. Bone loss during spaceflight causes a dramatic loss in bone tissue, approaching 2% bone density loss per month of exposure. Disuse can also cause a rapid loss of bone similar to that seen during spaceflight. These observed gains and losses of bone under varying mechanical loads indicate that strain plays a key role in regulating bone density.

Goodpasture's syndrome

When vertebrates transitioned from water to land approximately 300 million years ago, it caused selection pressure for type IV collagen, which acts to physically stent the walls of the lung air sacs, or alveoli. We know from studies of Goodpasture's syndrome that the three isoforms of type IV collagen evolved sometime between the emergences of fish and amphibians through selection pressure for specific amino acid substitutions that rendered it more hydrophobic and negatively charged, physically preventing the exudation of water and proteins from the microcirculation into the alveolar space.

Goodpasture's syndrome is an autoimmune disease causing simultaneous kidney and lung failure due to pathogenic circulating autoantibodies targeted to a set of discontinuous epitope sequences within the noncollagenous domain 1 (NC1) of the three chains of type IV collagen [3(IV)NC1], referred to as the "Goodpasture autoantigen." Basement membrane-extracted NC1 domain preparations from *Caenorhabditis elegans*, *Drosophila melanogaster*, and *Danio rerio* do not bind Goodpasture autoantibodies, while frog, chicken, mouse, and human 3(IV)NC1 domains bind autoantibodies. The 3(IV) chain is not present in worms (*Caenorhabditis elegans*) or flies (*Drosophila melanogaster*), and was first detected in fish (*Danio rerio*). Importantly, native *Danio rerio* 3(IV)NC1 does not bind Goodpasture autoantibodies. Unlike the recombinant human 3(IV)NC1 domain, there is complete absence of autoantibody binding to recombinant *Danio rerio* 3(IV)NC1. Three-dimensional molecular modeling of the human NC1 domain suggests that evolutionary alteration of electrostatic charge and polarity due to the emergence of critical serine, aspartic acid, and lysine amino acid residues, accompanied by the loss of asparagine and glutamine, contributes to the emergence of the two major Goodpasture epitopes on the human (IV)NC1 domain, as it evolved from *Danio rerio* over 450 million years. The evolved 3(IV)NC1 domain is a natural physicochemical barrier against the spilling of serum and proteins from the circulation into the alveoli or glomeruli, due to its hydrophobic and electrostatic properties, respectively, which were more than likely the molecular selection pressure for the evolution of this protein, given the oncotic and physical pressures on the evolving barriers of both the lung and kidney.

Salinity and the vitamin D receptor

Speculation that external selection pressure has primarily been on the lung innate host defense system is further reinforced by the following example of external selection pressure affecting molecular lung development and evolution. Increasing salt content of the oceans, possibly as a result of the desiccation of watery environments, is thought to have driven vertebrates onto land (the Romer

Hypothesis), displayed as a geochemical selection pressure for lung evolution. This hypothesized mechanism of action for the effect of salinity resulted from its inhibition of cAMP bioactivity, counterbalanced by increased VDR, which independently stimulates tissue-specific innate host defense, since fish exposed to elevated salinity show coordinated, adaptive changes in both vitamin D hydroxylation and the corresponding VDR. This selection pressure for host defense may, in turn, have referred back to the anatomy of the physostomous fish swim bladder, which is connected to the gut tube by the pneumatic duct, allowing the fish to inflate its swim bladder by rising to the surface and taking a bite of air shortly after hatching. This adaptation to go to the surface for air may have been the antecedent of the transition by this class of fish from water to land. It may also have produced the positive selection pressure for host defense, since bacteria could have entered the swim bladder via the pneumatic duct. In support of this concept, leptin produced by the alveolar lipofibroblasts, and lipopolysaccharide, the major component of the outer membrane of Gram-negative bacteria, have similar effects on type II cell differentiation, suggesting that the constitutive evolutionary selection pressure may have been for host defense and that leptin evolved as the constituent mechanism for regulated host defense and barrier function. Such preadaptive balancing selection for the regulation of innate host defense in combination with vitamin D regulation of blood pressure may have resulted in the genetic shift in the P450 cytochrome CYP3A from a salt-sensitive to a salt-resistant form, facilitating the migration of humans away from the equator. It has been discovered that the salt-resistant form of CYP3A is among genes that have evolved in humans recently and rapidly. This may have been due to domestic salt use by early humans, unintentionally causing both more autoimmune diseases due to the inhibition of antimicrobial peptides and elevated blood pressure due to the maladaptive response to salt. Deleting both VDR and 1α-hydroxylase activates vitamin D in mice, causing elevated blood pressure, which supports these observations. The effect was normalized by feeding the animals vitamin D. In other such studies, the immune response was similarly affected.

Proximately, the result of such positive selection pressure for vitamin D metabolism may have been the northerly exodus of progressively lighter-skinned people out of Africa. Distally, the result may have been the acquisition of autoimmune diseases and the propensity to elevated blood pressure; for each 10° north or south of the equator, blood pressure increases by 2.5 mmHg (millimeters of mercury) and hypertension prevalence increases by 2.5%. As an aside, a literature search for vitamin D and blood pressure yielded 888 hits, and for immunity, yielded 1186 hits.

This phenomenology is consistent with the racial and ethnic incidences of hypertension, which may be linked to the evolutionary significance of other cytochrome P450 genes that cause disease susceptibility or resistance. For example, CYP3A may represent the most important P450 gene family, because of the wide substrate specificity of these enzymes, induction by many commonly used drugs, and its high level of expression in the liver and gastrointestinal tract. This developmentally evolutionary population genetic approach reflects the robust power of the cell-centric understanding for the First Principles of Physiology.

References

Aris-Brosou, S., Chen, X., Perry, S.F., Moon, T.W., 2009. Timing of the functional diversification of alpha- and beta-adrenoceptors in fish and other vertebrates. Ann. N.Y. Acad. Sci. 1163, 343–347.

Bridgham, J.T., Carroll, S.M., Thornton, J.W., 2006. Evolution of hormone-receptor complexity by molecular exploitation. Science 312, 97–101.

Eales, J.G., 1997. Iodine metabolism and thyroid-related functions in organisms lacking thyroid follicles: are thyroid hormones also vitamins? Proc. Soc. Exp. Biol. Med. 214, 302–317.

Torday, J.S., Rehan, V.K., 2002. Stretch-stimulated surfactant synthesis is coordinated by the paracrine actions of PTHrP and leptin. Am. J. Physiol. Lung Cell. Mol. Physiol. 283, L130–L135.

CHAPTER 10

The atom and cell are homologous point sources in a quantum world

Introduction

The perception of time as unidirectional and asymmetric is consistent with our intuitive sense of biological time, since we humans witness the linear trajectory of our lives and that of those around us. However, this seemingly materially based impression may be a biological illusion. The rapidly emerging science of Epigenomics encourages us to reappraise the significance of the previously mysterious and obligatory return of all eukaryotic organisms to the unicellular zygotic phase. In conventional terms, the zygotic unicell that results from the fertilization of two gametes following meiotic division has always been assumed to contain all of the genetic information necessary to form a new individual. As the impact of epigenetics has continually enlarged, it can now be said that the recapitulation of all multicellular eukaryotic organisms through the unicellular zygotic state serves a further crucial role—it is during this stage that there is critical adjustment and modulation of a stream of microepiphenomena experienced within the parental macro form. It performs this action as a critical recentering phase oriented toward cellular principles and imperatives. As a result, the unicellular zygotic phase can be considered the actual epicenter of eukaryotic life. In such terms, the macroorganic phase is in service to the zygotic unicell, missioned with acquiring epigenetic environmental novelties. The zygotic unicellular phase represents a critical transitional moment when epigenetic marks that have been acquired in the macro phase are either expressed during further development or discarded. The significance of this regulatory mechanism becomes clearer when the actual circumstances of multicellular

Quantum Mechanics, Cell-Cell Signaling, and Evolution
https://doi.org/10.1016/B978-0-323-91297-6.00014-X

eukaryotic life are considered. Eukaryotic macroorganisms are multicellular units comprised by an integrated collaboration of innate cells and a dominant fraction of obligatory and facultative microbial life that is both mutually collaborative and competitive.

Under conditions in which epigenetic impacts are recognized as substantial, target organisms that are seen as ensembles of linked and co-dependent life experiences, environmental impacts are sources of ambiguity among the co-aligned constituents. It is suggested that a means toward understanding epigenetic impacts in these terms is reconciled by the settlement of these environmental ambiguities as quantum phenomena. The centrality of this quantum action is at the level of the zygotic unicell, due to its unique means of synchronically and diachronically negotiating the space-time of physiology.

The Quantum Physicists Einstein, Feynman, and Bohm all thought that time was an artifact. Einstein concluded that the past, present, and future all exist simultaneously. In his book "Relativity," Einstein said that "Since there exists in this four dimensional structure (space-time) no longer any sections which represent 'now' objectively, the concepts of happening and becoming are indeed not completely suspended, but yet complicated. It appears therefore more natural to think of physical reality as a four-dimensional existence, instead of, as hitherto, the evolution of a three dimensional existence." Perhaps this novel view of biology will be validated by such physics, and vice versa.

Einstein proved that time is relative, not the absolute classical Newtonian mechanics that is generally assumed. However, most physicists have been reluctant to let go of any ordinary assumptions that are made about time. Physicists other than Einstein reached similar conclusions about the relative nature of time and even made dramatic advances toward a timeless perspective for the Universe. Yet they also were unable to change the time-based mentality ingrained in mainstream physics, and in society at large. Richard Feynman developed the most effective, explanatory interpretation of Quantum Mechanics, known today as the "Sum Over Histories" (SOH), describing time simply as a direction in space. The SOH stipulates that the probability of an event is determined by summing together all the histories of that event. A particle traveling through space can move in many different ways—curved, oscillating, squiggly—and either backward or forward on time paths. All that is left of these pathways when they are summed up as vectors are those that ascribe to the laws and forces of Nature. In other words, SOH is consistent with conventional "clock" time direction, a simple path in space that is more probable than any more abstract direction time might have taken otherwise.

David Bohm expresses the idea that there are two realities in his book "Wholeness and the Implicate Order," namely, the explicate and implicate. The explicate is the way in which we conventionally think of reality through our evolved subjective senses. Reality lies in the Implicate Order, which he describes as a singularly continuous field, where time is similarly nonexistent. Within this system, other worlds are just other directions in space, some less probable, others equally as probable as the direction we experience. In "Cosmology from the Top Down," Stephen Hawking states that "Some people make a great mystery of the multiuniverse, or the Many-Worlds interpretation of quantum theory, but these are just different expressions of the Feynman path integral." Darwinian evolution makes no room for such ambiguity; in contrast to that, Quantum Mechanics melds with biology when the zygote is viewed as the focal point of the life cycle.

Biology and physics

There is a fundamental problem in reconciling the physicist's view of time and that of the biologist, which came to light in a public debate between Einstein and Henri Bergson in Paris on April 6, 1922. Einstein favored the absence of time in the Universe, whereas Bergson was adamant about the need for time in human psychology. Bergson's biologic perspective was subjective, but it was more comforting as it conformed to comfortable, entrenched, intuitive associations and correlations. On the other hand, Einstein's view was objective, based on mathematics.

Similarly, there has been a significant shift against the widely held belief that all important genetic change is random mutational variation. This viewpoint has yielded to the contemporary understanding of the Epigenome, allowing for a similar shift in attitude from classical Newtonian physics toward the quantum world. Such a recalibration is required since the functioning genetic complement of any multicellular organism is an ongoing, dynamic interrelationship between any species' genome and an agitating epigenetic realm with all its diverse inputs.

Our current perception of the balance between intrinsic and extrinsic factors and their differing hereditary modes has also changed. The essential character of epigenetic influence is now acknowledged throughout evolutionary development. As a result, there is both mandate and opportunity to critically evaluate where the central control of this empowered Epigenome lies. By merging epigenetics with complex physiology, the unicell is that center, transmuting space-time by simultaneously knowing past, present, and future.

Two aspects of evolutionary development are essential for recognizing the primacy of the zygote in epigenetic inheritance: First, genetic assimilation into a species genome and its ultimate expression obviously differ and proceed in a more complex manner than the seemingly direct mechanism of uncovering certain morphologic features by selective breeding; second, there is now an opportunity to consider that the expression of genetic capacity is exerted through changes in both the intrinsic genome of an organism and its epigenome as quantum phenomena. Under such circumstances, our understanding of epigenetic action is considered a "field of probabilities" that might eventually be scrupulously modeled using quantum statistical mechanics rather than a general quantum frame that lacks similar rigor.

The issue of the discontinuities of evolution had been of particular interest to Eldredge and Gould decades ago. They referred to those discontinuities as "Punctuated Equilibrium." On the other hand, Darwin thought that evolution was a gradual process. Müller and Newman propose emphasizing the dynamics of developmental systems instead of natural selection, complementing adaptation with emergence and contingency with inherency. This controversy is like that over whether light is a particle or a wave. Bohr resolved that issue by pointing out that the difference is due to how you measure it, referred to as his "Complementarity concept," because in reality evolution is both gradual and punctuated, the cell being the ultimate determinant.

Any epigenetic influence is either active or inactive, depending on its status as a newly integrated part of an organism. Goswami addresses genetic variations in a similar vein since they derive from the same quantum energy state. But given the long-standing stability of some species, extending over millions of years, there must be some way to inactivate epigenetic inheritance, and even intrinsic genetic mutations. So the question is "Where does the regulatory agent reside within eukaryotic organisms to achieve such outcomes?" The zygote is that agent, determining the epigenetic genome, dictating the possibilities implicit in any of many available environmental states. That agency can be best understood as a quantum phenomenon.

Goldenfield has said that viewing evolution as a subset of population genetics is artificially limiting. Consequently, the way in which evolution links to the environment has not been satisfactorily explained. It is proposed that evolution should be understood as a problem of nonequilibrium statistical mechanics of genetic elements, which recalls Goswami's call for reconceptualizing evolution in quantum terms—he contends that variation occurs at the genetic level, but that natural selection acts at the morphologic level as a macro event that is never one to one.

Kauffman and Gare have suggested that nature is simultaneously observing and observed in the process of becoming, bypassing classical physics in a world of actual events that happen whether or not we observe them. They propose that evolution must be reconsidered as always coexisting adjacent to other possible states of being, continuously constituting new boundary conditions that impose and release constraints—such contingencies are equivalents of the superimposition of states at the quantum level; a regulatory apparatus is implicit as the means for the zygote to operate and succeed.

The unicell can express or prevent the acquisition of epigenetic marks as a complex mechanism that includes the role of either sperm or egg. Mediators of this mechanism have been identified, such as 5-methylcytosine, which has been shown to program parental

DNA methylomes in the zygote. The DNA methylomes of either the egg or sperm differ significantly, providing a mechanistic basis for transgenerational inheritance. The programming of such transitions must emanate from the unicellular state. These epigenetic processes are contingent on such properties as imprinting and X chromosome inactivation. In this context, it is well known that the cells of the early mammalian embryo are diverse and subject to dynamic and heterogeneous epigenetic influences, and that embryonic stem cells and primordial germ cells are epigenetically modified by transcriptional circuits that will largely determine cellular fate. Epigenetic modifications present early in embryonic development are derivative of the zygotic unicellular phase as its progenitor. Consequently, genetic and epigenetic mechanisms that govern the transition from the totipotent zygote to diverse stem cell pluripotency within the embryo are regulated by conditions derived from the prior stage.

Recent mathematical modeling suggests that the unicell is controlled by a quantum-like master algorithm for the informatics of environmental pressure from the Epigenome by adjusting epigenetic marks considered to be within the frame of the collapse of the superimposition of states. This process has been compared with a photon whose wave and particle dimensions depend on context, yet are always exhibiting both capacities, the history of a photon being in its simultaneous multiple chronologies, all entangled as if all had been experienced. Similarly, the zygote is the sum of its chronologies, always representing more than its current physical form, such phenomena having superseded our conventional notion of space-time, a conserved epigenetic mark acting to advance a zygote as both a future implicate and one of many present forms of a contextually dependent explicate. That mark may or may not be perpetually expressed, contingent on future situations since both states exist simultaneously, past and future alike—contrary to any simple arrow of time, its future can settle its past in a continual state of reciprocity, future and past coexisting, one state collapsing into expression in a manner that simultaneously reworks its past. Both states of being exist simultaneously as implicate and explicate, dependent on a future state, reflective of its Quantum Mechanical origin.

The status of the unicell manifests its macroorganic form outwardly, permitting the superposition of latent states to collapse into that best suited for survival. It manages to do so by affecting current epigenetic marks, transforming them simultaneously into both historical and latent future heritable forms. The consequent form exists as both present and future based on homeostatic parameters, which are dynamic variables contingent on both current environmental and long-range geologic terms. In this way, organisms successfully channel themselves within homeostatic boundaries for the present, and at the same time are fully enabled to cope with widely shifting environments, always referencing the First Principles of Physiology.

The counterintuitive capacity of quantum states to enable past and future has been demonstrated experimentally. Study of helium atoms shot through templates created by lasers has shown that the condition of the second grate affects how the atom behaved as it had passed through the first grate. It is feasible for a moving object to exist in two states at once in the quantum world, as both a particle and a wave, but it is not possible to accurately measure it in both states simultaneously. A research group in Australia has demonstrated that whether behavior is wave- or particle-like depends on the method of measurement; it is contextually determined by observer status. Something is only real if it is observed, so the past state can be reinterpreted by a future reality. Entanglement in which future actions may influence past events has been exhibited by photon pairs. Experimentally, it has been shown that if two pairs of photons are entangled and then separated, the status of the photons in the initial pairings now depends on the entanglement status of the two photons in the original pairs; fate of the first set of separated photons can then be altered even after they have been previously measured, changed, or even destroyed. In Quantum Mechanics, the future affects the past, both of which are part of a predictive and probabilistic indeterminacy.

Quantum Mechanics provide ways of better understanding biologic processes. For example, the actualization of epigenetic possibilities is due to collapse of the superposition of possibilities. This is the role of the zygote exerting itself through quantum systems as nonlocal correlations through entanglement at a distance and quantum coherence. McFadden and Al-Kahalili have discovered that bird navigation uses magnetic fields based on Quantum Entanglement, as do Monarch butterflies and fruit flies. Mae-Wan Ho had long held that the cell itself is a quantum unit.

When organisms experience environmental stresses, their response is distributed over a range of potentials, as is the case for newly acquired epigenetic marks. A particular stress might have two potential outcomes, whereas another stress might have two dissimilar responses—the individual organism bears all of these potential responses and may potentially use any of the four types of outcomes. The zygote collapses this superposition of possible outcomes idiosyncratically, formulating its own past. In essence, the unicell is deciding its past based on one set of collapsed memories versus the other. In this way, the zygote is producing a space-time shift, allowing the individual to gain from its future by forming a better past through the collapse of the superposition of differing potential states.

A similar circumstance has been identified in dreaming. Whether we dream in color, black and white, or both has been investigated—we dream in both color and black and white. Seniors dream more in black and white. A study was conducted, hypothesizing that memory plays a role in whether people say they dream in black and white or in color. The findings were that remembering dreaming in color is related to when the dream is recalled and how frequently such recollections occur. Recollection of dreams is subjective and depends on context. On an individual basis, dreams may be reported to be one or the other, or neither. Dreams are brainwaves that manifest both in color and in black and white at the time they are experienced. The individual becomes an observer of their own prior experience when the individual recalls the dream, making the determination as to whether it was actually dreamed in color or in black and white. That dream then becomes a memory of a past event, solidified as one or the other, the active past having existed as a superimposition of states. Experiencing that dream upon awakening as one type or another, it collapses into memory and is stored as past history, upon which future actions can be taken. The present observation is effectively making the past shape future actions, the past action having been decided on the basis of an ex post facto occurrence. This is a form of entanglement that is demonstrative of the determination of the actual past by a future entangled quantum event within the zygote.

This reprise of evolutionary development, placing the zygote at the epicenter of a quantum framework, is advantageous in evaluating inputs and their activation or suppression to identify and research where they operate. That demands greater precision of terms and methods of quantum statistical mechanics. A series of articles on evolution from unicellular to multicellular organisms based on cell-cell communication as a varied, integrated approach based on experimental data has been offered.

The relationship between space-time and environmental and epigenetic effects is an imaginative cellular-molecular process. This is particularly true when seen through these deep physiologic pathways. Equally important, the cellular-molecular approach to evolution from its origins eliminates the conventional view of time and space in understanding evolution.

On the origin of complex physiology

Considering the biologic relativity of time to be fundamental to evolutionary development, the origin of complex physiologic traits must be constrained by that narrative. If the cellular-molecular mechanisms that have given rise to complex physiologic traits follow on from the present back to the protocell, the temporo-spatial aspects of physiologic mechanisms are epiphenomena based on the same boundaries that have just been enumerated. In the first book based on this approach, entitled "Evolutionary Biology, Cell-Cell Communication and Complex Disease" (Torday and Rehan, 2012), cholesterol was used as a cipher to understand evolution from its primordial origins, since cholesterol requires 11 atoms of oxygen to synthesize one molecule of cholesterol, highlighting the requirement for atmospheric oxygen, which arose beginning in the Phanerozoic Era, increasing and decreasing consequentially several times over the past 500 million years.

The emergence of cholesterol was critically important for the evolution of eukaryotes, the presence of cholesterol in the cell membrane promoting metabolism, locomotion, and respiration by allowing for a thinning of the cell membrane's phospholipid bilayer, giving rise to all of the enumerated traits. Consequently, cholesterol in the cell membrane facilitated cell-cell signaling by soluble growth factors that are secreted by one cell type and to another nearby cell with a receptor embedded in such a lipid raft. Through this means, receptor-mediated cell-cell signaling ultimately gave rise to the endocrine system.

Evidence for the causal, functional relationship between cholesterol and eukaryotic cellular evolution is seen with respect to phylogeny and development as histories for the role of cholesterol in the respiratory and neuroendocrine systems. Cholesterol is the simplest lung surfactant that mediated the evolution of the vertebrate gas exchanger. Cholesterol is secreted by the fish swim bladder epithelial cell lining, preventing self-adherence of the walls of the swim bladder. Cholesterol is a constituent of lung surfactant in subsequent forms of vertebrates associated with the water-land transition—amphibians, reptiles, mammals, and birds. The cellular-molecular evolution of the alveolus can be traced from the swim bladder of fish to the mammalian lung, by following the structural-functional interrelationships between epithelial-mesenchymal cell-cell interactions governed by soluble growth factors and the composition and regulation of surfactant production by the alveoli. Those interrelationships are causal because effective surfactant production is necessary for the stepwise increase in the efficiency of gas exchange from amphibians to reptiles, birds, and mammals. The only way in which gas exchange can become more efficient biologically is by reducing the alveolar diameter, increasing the surface-area-to-blood-volume ratio, thus enhancing gas exchange.

As evolutionary selection pressure for reduction in the alveolar diameter increased, the only way to prevent alveolar collapse due to increased surface tension based on the Law of Laplace was to produce either more lung surfactant or more effective lung surfactant. This action counters the otherwise ever-increasing surface tension caused by the evolutionary positive selection pressure for the decrease in alveolar diameter, always threatening the collapse of the alveoli. Cholesterol in the plasma membrane, which first facilitated oxygenation in unicellular eukaryotic organisms, is the functional analog of the surfactant facilitating oxygenation in the lung and swim bladder.

The pinnacle of lung evolution in mammals is the stretch regulation of lung surfactant, providing *ad hoc* control for this critical property of the alveolus. Distension of the alveolus coordinately increases the expression of Parathyroid Hormone-related Protein (PTHrP) by the alveolar type 2 cell with leptin signaling by the neighboring lipofibroblast. This coordinated evolutionary adaptation confers optimal dynamic regulation on surfactant production, preventing life-threatening atelectasis.

From a cellular-molecular evolutionary standpoint, it offers an ontogenetic/phylogenetic link that refers all the way back to the unicell and the First Principles of Physiology. These interchanges are selective reciprocations between environmental stresses and the organism, epigenetic implicates interconnecting with specific explicates. No way could such complex processes have evolved absent the malleability within the system of quantum inference for the best long-term environmental outcomes, which is the specific regulatory domain of the zygote.

Surfactant must be produced at birth to prevent alveolar collapse over the short-term ontogenetic history of the organism. If Scap-1, which is necessary for cholesterol synthesis, is strategically deleted from alveolar epithelial type 2 cells (ATIIs), the lung compensates by forming more lipofibroblasts in the alveolar wall, making up for the poorer quality of the surfactant surface activity due to the absence of cholesterol (Besnard et al., 2009). In essence, the lung is referring back to the lipofibroblasts it employed earlier in its phylogeny to protect the alveolar wall against oxidant injury.

In the above-mentioned study of Scap-1 deletion, the lipofibroblasts also expressed more Peroxisome Proliferator-Activated Receptor gamma (PPARγ). That refers back to the appearance of peroxisomes as a means of protecting against oxidant stress, which causes calcium release from the endoplasmic reticulum, risking cell death due to excess cytoplasmic calcium. The cell copes with this by either elaborating or acquiring the peroxisome, which uses neutral lipids to buffer the excess calcium. Lipofibroblast differentiation is determined by PTHrP, produced by the neighboring ATIIs; in turn, the lipofibroblast produces leptin, which acts retrogradely to stimulate surfactant synthesis by the ATIIs. This paracrine mechanism is integrated by receptors on the surface of the targeted cell type, mediating growth factor signaling, producing second messengers that induce gene expression for lung development and homeostasis. Therefore, at all stages, reciprocations between the organism and its environment act as a continuing set of implicates and explicates to resolve physiologic stresses. The zygote embodies that regulatory process, determining those factors that will continue forward to govern embryological development.

The evolution of visceral organs

The molecular mechanisms that have facilitated lung evolution are homologs of other visceral organs based on their phylogenetic and ontogenetic origins. This is exemplified by PTHrP signaling, which is fundamental to the development and physiologic homeostatic control of the skin, kidney, and skeleton. PTHrP mediates electrolyte and fluid homeostasis within the glomerulus, skin barrier formation, and skeletal plasticity. The causal basis for these interrelationships is demonstrated by deleting the PTHrP gene from mouse embryos, resulting in failed formation of lung alveoli, kidney barrier function, and skin barrier function. The evolutionary origin of this physiologic complex was revealed by the observation that the PTHrP receptor gene duplicated during the water-land transition (Pinheiro et al., 2012). The physiologic stress caused by having to adapt to land caused vascular hypertension, precipitating shear stress specifically in those microcirculatory beds most affected—namely, the lung and kidney. Duplicating the PTHrP receptor gene amplified its signaling, particularly as a physiologic receptor. That caused increased numbers of alveoli since PTHrP promotes both epithelial and endothelial cell proliferation. The primitive fish glomus would have homologously increased its PTHrP signaling between the podocytes and mesangial cells lining it, as occurs during mammalian fetal development, the lung and kidney both functioning to produce amniotic fluid in the process of their mutual development. Hence, developmentally, phylogenetically, and evolutionarily, this process results in expansion of the alveolar and glomerular surfaces, facilitating fluid and electrolyte transfer in both structures, enhancing control of homeostasis for land adaptation. In the skeleton, increased PTHrP signaling due to the greater effect of gravity compared to buoyancy in water would have enhanced its bone plasticity, known as Wolff's Law, allowing for the five documented bone remodelings that occurred during the water-land transition, which are in the fossil record (Clack, 2012).

Mechano-transduction and evolutionary fundaments

As discussed above, lung surfactant is a stretch-regulated mechanism ensuring that when the lung expands, there is adequate surface activity to prevent the alveoli from collapsing on end-expiration. This process is coordinately regulated by the cell-cell signaling partners PTHrP and leptin. PTHrP is produced by ATIIs, and leptin is produced by lipofibroblasts. Their receptors reside on the apposing cells. When the alveolar wall is distended, these two ligands and their receptors are coordinately increased. This mechanism is homologous with the fish swim bladder, which uses surfactant to lubricate the inner surface of the bladder, preventing self-adherence of the bladder wall, ensuring efficient function. The swim bladder evolved in adaptation to gravity, mediating buoyancy in water by inflating or deflating to maximize the efficiency of swimming and feeding activities. The role of gravity in alveolar homeostatic regulation by PTHrP was experimentally determined by putting isolated ATIIs in free-fall using a rotating wall vessel bioreactor to mimic microgravity, demonstrating that under microgravity-like conditions expression of the gene for PTHrP decreased; the PTHrP receptor, leptin, and the leptin receptor are all also stretch-regulated, likely having evolved through positive selection for stretch-regulated surfactant adaptation on land.

Further understanding of the significance of mechano-transduction derives from microgravity-like exposure of yeast, the most primitive unicellular eukaryotes. Under such conditions, yeast lose their capacity to polarize and to bud. Lack of polarization is indicative of loss of calcium flux; inability to bud is the mechanism of reproduction in these organisms. Loss of gravitational tension on the cytoskeleton has profound effects on the physiologic status of the organism, reflecting its fundamental determination of the status of the cell as either homeostatic, mitotic, or meiotic. For example, under nutrient stress, the free-swimming amoeboid form of the slime mold *Dictyostelium* reverts to its colonial form, mediated by Target of Rapamycin (TOR) signaling to AKT. Atavistic traits referencing single-celled organisms have been exapted as complex physiologic traits like alveolar homeostasis.

Therefore, such complex physiologic arcs as that of cholesterol or the reciprocating, coordinate nature of mechano-transduction between environmental stresses and cellular structure-function reveal the integrated synchronic and diachronic basis for the mechanism of evolution. The evolutionary process can now be recognized as self-organizing and self-referential permutations and combinations of the unicellular state. It is this particular set of properties, in combination with contextual epigenetic environmental impacts, which forms the basis for biology. In this light, the transgenerational nature of epigenetic inheritance can now be understood from the perspective of the iterative return to the unicellular state. Therefore, rather than superficially recapitulating evolutionary processes, targets for research become readily available.

During meiotic reduction division, newly acquired epigenetic marks are assimilated. Those marks that are incompatible with development and homeostasis are rejected during morphogenesis as a result of embryonic lethality. However, the workings of the self-referential unicell and its environmental responsiveness as a quantum agency render selection no longer random. Moreover, the compartmentalization of the life cycle as infancy, childhood, adolescence, adulthood, or senescence serves to regiment further expression of acquired epigenetic marks. This is an interactively integrated mechanism, since epigenetic marks affect the endocrine system, which determines the duration of the phases of the life cycle, as well as the organism's behavior. This reciprocally balanced iterative process prevails over evolution.

The role of physiologic stress in causing mutations and duplications was alluded to above. Such injury-repair processes are not random; they are bounded by the structural-functional cellular niches in which reactive oxygen species are generated by microvascular shear stress, favoring those structures and functions that alleviated such stresses over time. The fundamental unicell remains at equipoise as an iterative quantum agent, by which evolution takes on a mechanistic explanation that supersedes random mutation and natural selection.

Such speculation is supported by the three gene duplications known to have occurred during the water-land transition about 300–500 million years ago, namely, for the Parathyroid Hormone-related Protein Receptor (PTHrPR), the Glucocorticoid Receptor (GR), and the βAdrenergic Receptor (βAR); these duplications were accompanied by the mutation of the Mineralocorticoid Receptor to form the Glucocorticoid Receptor. Note that all of these gene duplications targeted receptors, which is unlikely to have been a coincidence. It was more likely due to the nature of the mechanisms involved. *A priori*, it is more efficient to duplicate a receptor, which is a natural amplifier. *A posteriori*, the specific physiologic changes were for a series of receptors that signaled through a series of highly evolved downstream second messengers and their strategically important phenotypes, without disruption of the evolved homeostatic control mechanisms.

Norman Horowitz had described a similar mechanism (Horowitz, 1945), speculating that if an evolving organism used up all of a given nutrient in its niche, it evolved the next enzyme in the metabolic pathway upstream, and so on, and so forth. But he did not provide a mechanistic test of such events as those above that emanate from the homeostatic constraints on the system, from its unicellular beginnings.

Such a process involving receptor duplication is consistent with the observation of Terminal Addition, first proposed by Ernst Haeckel in defending his Biogenetic Law. Terminal Addition refers to the appearance of a new trait at the end of a chain of interrelated traits. However, no particular mechanism for this property has previously been speculated.

In further support of this hypothesis, another gene duplication that occurred during the water-land transition was the Goodpasture syndrome type IV collagen isomer. It is expressed in alveoli and glomeruli, first appearing in amphibians over the course of vertebrate evolution. This form of type IV collagen was advantageous in facilitating the water-land transition because it is more hydrophobic than the other forms of type IV collagen expressed by epithelia, prohibiting fluid and electrolyte loss across these barriers. However, some individuals succumb due to this isoform because they develop antibodies against it, causing an immune reaction resulting in renal-pulmonary failure. In contrast to the hormone receptor duplications mentioned above that evolved under homeostatic control, the Goodpasture type IV collagen isoform did not, explaining why the difference in the end results.

Any consideration of the origin of system-wide metabolic pathways or its evolutionary development of novelty involves any impulse toward self-organization that must be present for such end-points to be observed. An appropriate answer is the feedback stigmergic loop. The process of stigmergy was first described by Grasse and Parunak as a feedback system in which any action leaves some kind of trace in a medium. In biologic terms, communication leaves a trace of its action or path on the environment. This in turn produces a further action, either the entity leaving the first trace or others that follow, which then generates other marks or different traces. Heylighen defined stigmergy as "an indirect, mediated mechanism of coordination between actions, in which the trace of an action left on a medium stimulates the performance of a subsequent action."

The best studied example is termite mounds, and it is hypothesized that there is tandem activity at the cellular level. This biologic mechanism connects the causal train of physiologic events within an organ, or by providing the necessary links between organs, i.e., lung, kidney, brain, skin. Such gene duplications may be expressed in different environmental niches within any developing organism. In such instances, the individual constituents of any cellular ecology leave a stigmergic trail that enables self-organization among individuals, bearing in mind that the initiating factors for stigmergy can arise from primary impulses that need not relate to any later developments.

Biologic time as artifact

Einstein eliminated time as a fixed physical dimension with his Theory of Relativity, launching physics as a predictive science. On the other hand, time remains central to biology, which has inhibited progress in understanding evolution. Once physiologic structure and function are reduced to the unicellular state based on cell-cell communication, and it is understood that it is self-perpetuated in that context, the time dimension is superfluous in understanding the fundament of physiology.

A recent breakthrough in the transition from descriptive to mechanistic biology is the realization that epigenetic inheritance is critically important to the reproductive aspect of the evolutionary process. Lewis Wolpert has quipped that "It is not birth, death or marriage, but gastrulation which is truly the most important time in your life," since gastrulation is the stage of embryogenesis when the mesoderm is introduced between the endoderm and the ectoderm, mediating further development of the fetus through interactions between these germ layers. And it is now known that the transition from the blastula to the gastrula is under epigenetic influence, providing a mechanistic foundation for understanding how epigenetics affects embryogenesis. Thus, it is the immediate environment that in large part determines the immediate impact of epigenetic inheritance that is limited and edited during meiosis, and then during morphogenesis as a derivative of the zygotic state. Consequently, the zygote must be appreciated as the pivotal intermediary in mediating such epigenetic influences, not just because of the juxtaposition, but as the embodiment of the iteratively elaborated organism. Over the course of sexual reproduction, it is the reprising of the unicell that is the controlled process that facilitates that crucial recentering and modulation of both the epigenome and the genome alike. That recapitulating zygote provides the unique capacity for the stability and plasticity of species. That cannot be readily accomplished by mere random mutation and selection. In fact, the optimal perspective on the role of the zygote is to see beyond the structure and function of multicellular organisms, accepting that the organism never actually leaves its unicellular state. Those "tinkerings" that Francois Jacob described as evolution (Jacob, 1977) are the ways in which the unicell has adapted to an ever-changing environment. Essentially, the unicellular state delegates its progeny to interrogate the environment as its agent, obtaining data from its surroundings to forewarn the recapitulating unicell of ecological changes that are occurring. Through the acquisition and processing of epigenetic marks during meiosis, fertilization, and embryogenesis, and over the course of the life cycle, the unicell remains in synchrony with environmental changes.

In conclusion—The cell as the first niche construction

The Niche Construction Theory hypothesizes that organisms actively participate in shaping and generating their own immediate environments. Classically, the earthworm makes the soil surrounding it more conducive to an otherwise water-adapted organism, allowing it to retain its water-borne kidneys. Yet the incursion of any organism into the environment is the microscopic analog of niche construction, enacted at the cellular level. Hence, the zygote is an inextricable form of niche construction within its own discrete boundaries.

Niche construction is an iteratively modular process, and eukaryotic organisms can thereby functionally merge evolution and ecology into one continuous, harmonious process through successive steps. The participation of the zygote coordinates this process by continuously utilizing the epigenetic impacts acquired by the phenotype, the myriad heterogeneous biologic forms all deriving from the modulation of the unicellular state. It is the continuity of the zygote referencing the First Principles of Physiology as the manifestation of the Singularity that co-aligns biology with physics through an Einsteinian rejection of biologic time as fixed or even relevant.

Beyond theory, there is a means for this type of process to spill forward in biologic expression without outside direction through both internally and externally generating indirect cues. Stigmergy only demands that the participants be able to send and receive information and take action based upon it, and assumes that participating agents are individually goal directed, independent of the goal itself. Biologically, stigmergy sufficiently seeks to maintain its own homeostatic preferences. Since the individual participants will have independent goals in any mixed cellular ecology, there is a natural division of labor, and the variety of these participants working together builds complexity in sequence or in parallel, depending upon the continuous stream of information from within any ecological niche, or from without, based on impacting epiphenomena. The available information is sourced through direct communication and nondirected data in which the sender and receiver are not necessarily aware of one another. The inherent value of stigmergic systems lies in the direct coordination among individual players being unnecessary. In one sense, indirect information is just an artifact of sender/receiver units coexisting among one or more of the participants through the stigmergic feedback loop. It need not have anything to do with the intentionality of the sender. Coordination emerges from tissue ecologies directly and indirectly. If information is not expressly directed and received, it can become ambiguous. The consequent collapse of the superposition of possibilities settles through the nondirected use of information as noise to some, significant to others, leading toward an emergent resolution of the ambiguities reflected in the totality of the information field. Once resolved, some leave a trace of negative reinforcement and others become self-reinforcing. In this way, the superposition of possibilities is perpetually resolved through the ecosystem's participants experiencing various epiphenomena within their homeostatic milieu.

Since all multicellular eukaryotes are coalescing cellular and noncellular participants, each constituent of the organism experiences time within its own scope, within which biologic time becomes relative. Stigmergy can provide unity between quantum ambiguities in any information system, and their eventual collapse into biologic expression experienced as an iterative phenomenon throughout any cellular ecology provides the effective drive toward ecological self-organization.

The zygote is the confluence of the history of the organism and its character as a point source, both the observer and the observed, collapsing the superposition of select epigenomic states. That process yields dynamic biologic outcomes in continuous flux, but with the background of a stable genome. This unique cellular agency is where Bohm's realms of the explicate and implicate connect as the present, past, and future, melding physics and biology together.

References

Besnard, V., Wert, S.E., Stahlman, M.T., Postle, A.D., Xu, Y., Ikegami, M., Whitsett, J.A., 2009. Deletion of Scap in alveolar type II cells influences lung lipid homeostasis and identifies a compensatory role for pulmonary lipofibroblasts. J. Biol. Chem. 284, 4018–4030.
Clack, J., 2012. Gaining Ground. Indiana University Press, Bloomington.
Horowitz, N.H., 1945. On the evolution of biochemical syntheses. Proc. Natl. Acad. Sci. U. S. A. 31, 153–157.
Jacob, F., 1977. Evolution and tinkering. Science 196, 1161–1166.
Pinheiro, P.L., Cardoso, J.C., Power, D.M., Canário, A.V., 2012. Functional characterization and evolution of PTH/PTHrP receptors: insights from the chicken. BMC Evol. Biol. 12, 110.
Torday, J.S., Rehan, V.K., 2012. Evolutionary Biology, Cell-Cell Communication, and Complex Disease. Wiley, Hoboken.

CHAPTER 11

The Vitruvian Cell

Introduction

In his 2005 commencement speech at Kenyon College, David Foster Wallace related the "parable" about the two little fish:

> *There are these two young fish swimming along, and they happen to meet an older fish swimming the other way, who nods at them and says, 'Morning, boys. How's the water?' And the two young fish swim on for a bit, and then eventually one of them looks over at the other and goes, 'What the hell is water?'*

The following is an explanandum for "what the hell is water?" for we humans since we obviously are not immersed in water.

da Vinci's Vitruvian Man

da Vinci depicted his Vitruvian Man in 1490AD, his own human male form inside of a square, inside of a circle. It infers the interrelationship between Man (the square) and the Cosmos, with the square representing Bohm's Explicate Order and the circle representing his Implicate Order. The Explicate Order was derived from the Implicate Order when lipids formed a barrier between the two realms, with micelles spontaneously forming when lipid molecules were immersed in the primordial ocean. The micelles provided the space in the Implicate Order into and out of which particles freely could flow, some being entrained by Quantum Entanglement. Nonlocal entities like photons, electromagnetic fores, the moon, Pulsars affected the local particles within the cell, or "action at a distance," forming the basis for Symbiogenesis, the mechanism by which factors in the environment that posed an existential threat were internalized and made useful as what we refer to as "physiology."

Quantum Mechanics, Cell-Cell Signaling, and Evolution
https://doi.org/10.1016/B978-0-323-91297-6.00010-2

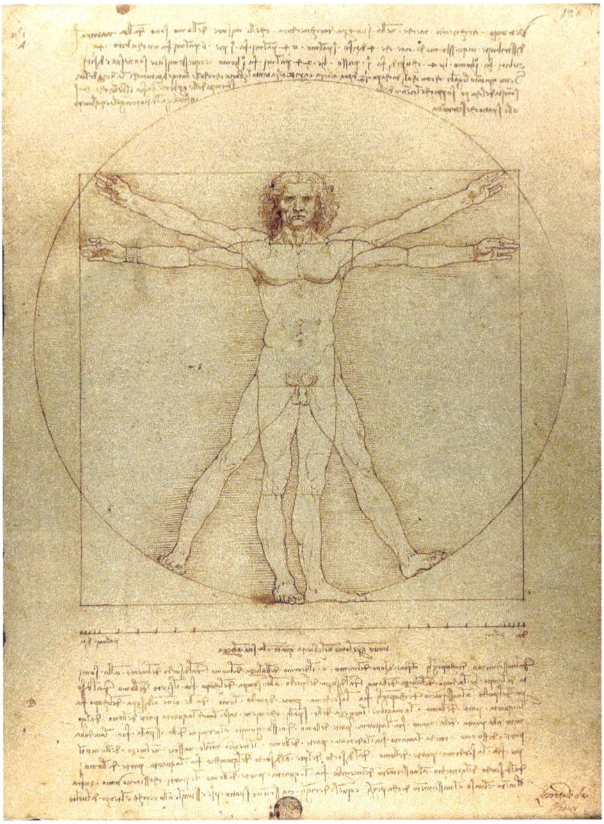

Vitruvian Man. da Vinci's depiction of Man circumscribed by a square and a circle. The text surrounding the image is Vitruvius's exhaustive measures of the human form.

But "squaring the circle," or forming a square with exactly the same area as a circle, has proven to be impossible based on the Lindemann-Weierstrass theorem because a circle is referred to as "transcendental," meaning it is not the root of any polynomial with rational coefficients, whereas a square is an algebraic irrational number. Perhaps the difference between the area of the square and that of the circle represents the "gap" between Bohm's Explicate and Implicate Orders? And since the formation of the Explicate Order represents the beginning of life, perhaps that gap is where life exists.

The text around the figure is Vitruvius's description of the metrics of the human form. It is reminiscent of the Greek Philosopher Protagoras of Abdera's aphorism that "man is the measure of all things." That, in turn, is consistent with the Anthropic Principle, that we are "in" this world, when in reality we are "of" it based on the Symbiogenesis Theory. The aggregate of that process is our physiology, which is where our mind/consciousness emanates from since we must comply with the same Laws of Nature as the Cosmos does.

Math, geometry, biology, and metaphysics

Nick Rossiter informs us that a square is nonexistent in Nature because the square root of two does not exist, yet we can perceive it in the Explicate Order. However, π is the ratio of the circumference of a circle to its diameter, reflecting the fact that it exists in the Implicate Order. It is interesting in this vein to consider that π is referred to as a "transcendent" number, meaning that it is nonalgebraic; it may also intuitively indicate that the circle references the origin of the cell in the Implicate Order, forming the Explicate Order by producing the lipid boundary between the two. The space between the square and the circle is what we measure inductively using the scientific method and approximate deductively through literature, poetry, art, economics, history, psychology, etc.

The circle is a mathematical homolog of the cell based on Knot Theory, proof of which lies in the ability to unknot the knots and reform a circle. As such, it exists within the Implicate Order, providing for its transcendent existence within the Explicate Order. Classically, the spherical cell is described as the optimal configuration for gas exchange based on its surface-area-to-volume ratio, but that is after-the-fact reasoning that does not lead to deeper understanding of the meaning of life. However, when seen from the perspective of its transcendent mathematical origin, it acts as the connection between the Implicate and Explicate Orders, accounting for both its ontology and epistemology. This difference between the function of the cell and its prima facie characteristics has been addressed by Etienne Roux, who has pointed out that defining physiology based on function is teleologic, though he didn't provide an alternative paradigm. The above provides such a paradigm.

Leonardo da Vinci

Leonardo da Vinci was on the right track, but he had no knowledge of the cell as the basis for human biology and how it literally forms the connection between life and the Cosmos. Briefly, the cell came into existence when lipid molecules were immersed in water, spontaneously forming micelles. Within those primitive "cells," particles would coalesce due to Quantum Entanglement, energized by gravity impinging on the curved surface of the micelle based on Einstein's Field Theory. Moreover, those particles were local phenomena that referenced nonlocal "events" out in the Cosmos. Among them was gravity, which we now know to be necessary for the formation of the cell itself and for its phenotypic identity. The influence of the moon and Pulsars are other nonlocal cosmologic effectors. So we are truly "authored" by the Cosmos.

Those Quantum Mechanical interrelationships formed the basis for Symbiogenesis. That iterative process gave rise to physiology as the compartmentalization and networking of such endogenized factors, accommodating the ever-changing environment. And because our internal milieu must ultimately comply with the Natural Laws of the Cosmos, it constitutes what we think of as mind/consciousness. Therefore, all of the aspects of life are homologs of our physiology.

Homology between the cell and a crossword puzzle

The way a crossword puzzle works, if you don't know "8-down," it "solves" itself by filling in the across spaces. Thought of as the overlap of the Implicate and Explicate Orders, the mind can work similarly to "solve" problems by conflating two or more ideas to circumscribe the answer. Other disciplines like science, literature, and poetry similarly juxtapose ideas and offer the opportunity to transcend the known to gain insight into what we don't know.

The converse is Andrew Crompton's interpretation of fonts, which become "invisible," blending in with the background due to their familiarity, like Husserl's explanation for the origin of geometry, which is like Platonic archetypes that are in our heads. So the idea is that we exist in that niche between the square and the circle in the

Vitruvian Man image, between the Implicate and Explicate Orders based on Bohm, or Heaven and Hell in humanistic terms. This niche began with the micelle providing a space within which Quantum Entanglement poised particles that were acted upon by nonlocal entities in the Cosmos, superseded by Symbiogenesis as the mechanism of evolution.

Ideally, this process functions optimally when there is free flow of energy through it from the Cosmos to the cell, to the organism, to the environment, to Gaia. That occurs when cells cooperate with one another metabolically, communicating by means of soluble growth factors and their cognate receptors. When a growth factor comes in contact with its receptor, it triggers the production of second messengers such as cyclic adenosine monophosphate and inositol phosphates, constituting an exchange of energy from one cell to another. Such energy cascades are under homeostatic control, maintaining cellular-molecular balance within and between cells. Under physiologic stress, cells may lose their homeostatic control, resulting in the production of reactive oxygen species (ROS). Such ROS can cause gene mutations and duplications in the course of re-establishing homeostatic balance; when homeostasis is re-established, the structure and function of the cells involved are said to have evolved. In this way, the cell and its surroundings remain at equipoise, even though the environment is ever-changing.

CHAPTER 12

Implicate and Explicate Orders as unconscious and conscious

We shall not cease from exploration
And the end of all our exploring
Will be to arrive where we started
And know the place for the first time.

"Little Gidding," T.S. Eliot

Introduction

Science asks us to suspend our common beliefs in favor of falsifiable evidence (Popper, 1959), and for good reason. We began life within an ambiguity of energetic states as negative entropy (Torday and Miller Jr, 2017) and have been making up "Just So Stories" to cope with that reality ever since.

But there is an ultimate Truth contained within the Cosmos that Physicist David Bohm describes as the Implicate Order (Bohm, 1980). We tend to universally express that concept as "something greater than ourselves," but the answer to that enigma has thus far remained intractable. However, scientific breakthroughs based on empiricism continue to encourage us to seek that ultimate Truth, such as Heliocentrism (Kuhn, 1957), the Periodic Table of Elements (Scerri, 2019), Evolutionary Biology (Darwin, 1859), Quantum Mechanics (Mehra and Rechenberg, 1982), and the Big Bang (Dodelson and Schmidt, 2020). And if there were some common origin from which all of them were to emanate, we should theoretically be able to identify it by finding self-similar patterns in each of them. That commonality is tentatively identified herein as the patterns formed by both the Periodic Table of Elements and Evolutionary Biology, originating from the way that gravity enables Quantum Entanglement

Quantum Mechanics, Cell-Cell Signaling, and Evolution
https://doi.org/10.1016/B978-0-323-91297-6.00008-4

(Torday, 2021)—when gravity strikes the curved surface of a protocell, it produces energy based on Einstein's Field Theory (Einstein, 1961). That energy sustains the Quantum Entanglement for the contents of the cell, in turn referencing the nonlocal gravitational force of the Cosmos.

Such hope for understanding life's mystery has been buoyed by the recent realization that the cell membrane behaves on two levels—as a barrier for the entry and exit of matter in and out of the cell and as a "Mobius strip" that accommodates the holistic nature of the cell as the conscience of the Cosmos (Torday, 2021). In that vein, it has been hypothesized that our physiology constitutes Bohm's Implicate Order, whereas our day-to-day behavior constitutes Bohm's Explicate Order. That way of understanding our state of being would explain the unconscious and conscious, for example, as the Implicate and Explicate Orders, respectively. And how and why under stress conditions, the two states merge together to one degree or another due to the influence of the endocrine system acting to mediate wave collapses due to the merging of the calcium fluxes within cells, neurons, and nonneurons alike (Torday, 2022). Such events as Near-Death Experiences, Out-of-Body Experiences, Maslow Peak Experiences, the Runner's High, and meditation may all be consequences of hormonally mediated wave collapses. Beyond that, the two-tiered dual nature of consciousness may explain the Young Double Slit Experiment (Young, 1804), the particle beam of light complying with the Explicate Order, the wave form with the Implicate Order. And the Libet Experiment (Libet et al., 1983) may be interpreted as experimental evidence for such a hypothetical duality of consciousness. The 300-ms time lag between sensing the electrical shock and the response may be due to our Implicate conscious response.

These insights have been revealed by an empiric approach to the evolution of physiology based on the principle of cell-cell communication, mediated by soluble growth factors and their receptors (Torday and Rehan, 2012, 2017). As such, this approach is unique among the many ways in which Evolutionary Biology has been addressed since it is solely based on empiric evidence, and is therefore faithful to Popperian testability and refutability (Popper, 1959).

The Periodic Table of Elements and You

Eric Scerri has written extensively about Mendeleev's Periodic Table of Elements. In his book "The Periodic Table: Its Story and Its Significance," Scerri (2019) importantly informs us that Mendeleev did not merely arrange the Elements based on their atomic number; he also used how they behaved chemically to produce specific salts to guide his decisions in constructing his version of the Periodic Table, taking into account variations like isotopic forms, reinforcing the diachronic across space-time use of atomic number to arrange the Table. Atomic number is a function of the number of protons in the nucleus of each Element, which references its diachronic origin.

Nucleosynthesis is the means by which stars produce light, in the process forming the Elements, arranged from the lightest to the heaviest (Smolin, 1999), from hydrogen to iron. That progression provides a "logic" for the Cosmos, and the process of Symbiogenesis (Margulis, 1993) provides the means by which organisms have coped with existential threats by endogenizing and compartmentalizing them to form physiology. Husserl tells us that this is the nature of geometry, for example, which exists within our physiology, but is manifested outward as lines and circles and triangles, not unlike Plato's archetypes. Such conscious properties are so ancient that we do not recognize their origins any longer. Such physiologic "elements" must conform to the same physical logic by obeying the Laws of Nature.

Such use of empirical knowledge is akin to the way that Evolutionary Biology has been reverse-engineered based on cell-cell communication mechanisms in developmental biology (Torday and Rehan, 2012, 2017), the three primary germ layers—endoderm, ectoderm, and mesoderm—interacting with one another sequentially, beginning with the fertilized egg or zygote, ultimately giving rise to the offspring. The premise is that embryology is the only mechanism we know of for generating form and function biologically. By superimposing the developmental cell-cell signaling mechanisms that determine form and function onto the phylogenetic changes in phenotype as speciation, using the same signaling mechanisms as a common denominator (Torday and Rehan, 2007), the underlying principles of change from the swim bladder to the lung, or the glomus of the fish kidney to the glomerulus of land animals emerge. And like the empiric details of the Periodic Table mentioned above, the hormonal effects on evolution provide the dynamic aspects of phenotypic change both ontogenetically and phylogenetically, independent of the time differences (Torday, 2022).

The inherent interrelationships between physics, chemistry, and biology are illustrated in Fig. 1. Subsequent to the Big Bang, depicted at the bottom of the figure, the Elements were formed through nucleosynthesis (Smolin, 1999) and distributed throughout the Cosmos. Organisms were able to cope with the environment by endogenizing physical factors that posed an existential threat. The Elements in the Periodic Table represent both their "synchronic" physical characteristics and their "diachronic" number of protons in their nuclei.

Similarly, evolution is constituted synchronically as embryologic development and diachronically as phylogeny. In this sense, the Periodic Table of Elements and the process of evolution are homologs, being derived from the same origin. This is consistent with the homology between the atom and the cell (Torday, 2018a), which

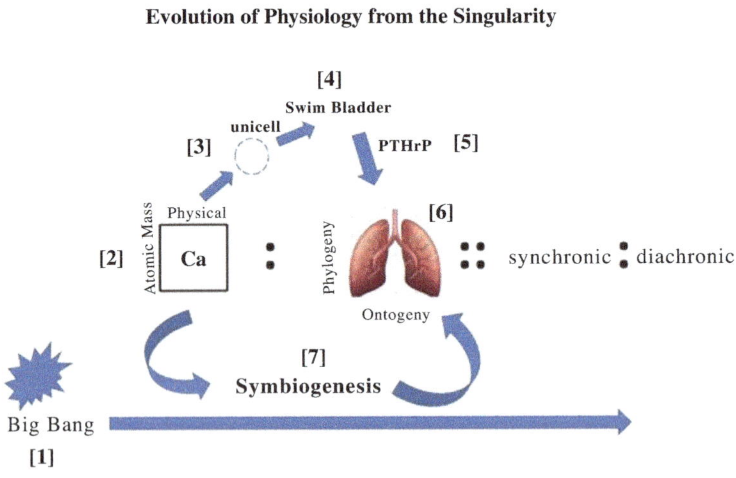

Fig. 1 Origin of physical, chemical, and biological homologies. Beginning with the Big Bang **[1]**, the Elements formed through nucleosynthesis **[2]** synchronically and diachronically as both physical characteristics and atomic mass. Unicells **[3]** gave rise to complicated physiology, such as the swim bladder **[4]**, PTHrP **[5]** facilitating evolution into the lung **[6]** mediated by symbiogenesis **[7]**. The Elements and biology are both synchronic and diachronic, indicating their homology **[from 2 to 6]**.

are both point sources that adhere to Pauli Exclusion Principle and Heisenberg Uncertainty Principle, conferring deterministic and probabilistic characteristics to each (Torday and Miller Jr., 2016b).

The Parathyroid Hormone-related Protein (PTHrP) epitomizes the cell-cell interactions that have mediated the evolution of vertebrate structure and function, as represented by the morphing of the fish swim bladder into the mammalian lung, the first 50 genes involved in the development of each being identical (Zheng et al., 2011).

These are conventionally thought of as time-based processes, but since Einstein tells us that time does not exist (Canales, 2016), these properties of biology exclusively reference space (Rowlands, 2015).

The question as to whether time exists in physics and biology or not has been discussed extensively. Einstein and Bergson debated this question publicly in 1922 (Canales, 2016). Because of the way that epigenetic inheritance works in service to the maintenance of the unicellular state (Torday, 2015a), its emerging prominence would favor the nonexistence of time in biology as an artifact of descriptive biology rather than mechanistic biology. The irrelevance of time in biology is consistent with the counterintuitive idea that the future determines the present (Di Corpo and Vannini, 2015). Our book (Guex et al., 2020) on "Morphogenesis, Environmental Stress and Reverse Evolution" provides empiric evidence for "reverse evolution," challenging the existence of chronological time in biology.

It is interesting to recall here the speculation of William Crookes in his 1886 Presidential Address to the Chemical Section of the British Association that a spiral representation of the Periodic Table (see Fig. 2) could be explained in terms of a progressive evolutionary genesis of the elements as a result of two forces, "operating in accordance with a continuous fall of temperature," and the other showing a sinusoidal variation (simple harmonic oscillator), connected with the electric force, together producing a (double helical) generation of elements of increasing atomic mass, but periodically similar chemical properties (Scerri, 2019). This speculation, which predated the discovery of basic atomic structure—the electron, the proton, and neutron—which predicted isotopes and accommodated the (undiscovered) inert gases, could now be seen as exemplifying the characteristic actions of

Fig. 2 Crookes "spiral" representation of the Periodic Table of Elements. He said that a spiral representation of the Periodic Table could be explained in terms of a progressive evolutionary genesis of the elements as a result of two forces, "operating in accordance with a continuous fall of temperature," and the other showing a sinusoidal variation (simple harmonic oscillator), connected with the electric force, together producing a (double helical) generation of elements of increasing atomic mass, but periodically similar chemical properties.

Nature when operating according to the (then unknown) universal "rewrite system" chronicled by Peter Rowlands in his book on "The Foundations of Physical Law" (Rowlands, 2015).

Thinking about how the dual forces of decreasing entropy and the sinusoidal oscillations/electrical force yielding a double helical generation of elements would coincide with the biology, such oscillations are reminiscent of the oscillating levels of oxygen in the atmosphere over the past 500 million years, fluctuating between 15% and 35% (Berner, 1999). The documented periodic increases in oxygen caused the widely recognized phenomenon of "gigantism" (Vermeij, 2016), whereas the physiologic effects of the periodic decreases in oxygen are not addressed anywhere in the scientific literature, other than what has been hypothesized regarding the evolution of endothermy (Torday, 2015b). Hypoxia is the most powerful natural physiologic stressor known. That correlates with specific physiologic changes that occurred in the pituitary-adrenal axis during this same epoch—the appearance of the Parathyroid Hormone-related Protein (PTHrP) gene in both the anterior pituitary (Mamillapalli and Wysolmerski, 2010) and adrenal cortex (Mazzocchi et al., 2001). The structural-functional effect of PTHrP is seen in the capillary arcades of the adrenal medulla (Wurtman, 2002), which expanded sometime during this era as well. That is physiologically relevant because the hormonal secretions of the adrenal cortex pass through the vascular arcades of the adrenal medulla on their way out of the adrenal gland to the systemic circulation. Since PTHrP stimulates the formation of capillaries (Diamond et al., 2006), the PTHrP produced in the adrenal cortex would consequently have increased the vasculature of the medulla. As a result of the above, the effect of the corticoids produced in the adrenal cortex passing through the adrenal medulla, stimulating the rate-limiting step in adrenalin production (Sharara-Chami et al., 2010), increasing adrenalin secretion by the medulla, would have been amplified by the increased surface area of the adrenal medullary vasculature (see above).

Adrenalin stimulates the production of lung surfactant by the alveoli (Lawson et al., 1978), increasing their distensibility and consequently their rate of oxygen uptake. So, in the aggregate, this cascade is primarily in service to alleviating the episodic hypoxia caused by the stepwise evolution of the nascent lung in adaptation to land in the short run. However, in the long run, PTHrP increases the formation of alveoli (Rubin et al., 2004), constitutively increasing oxygenation.

As for why that all may have occurred, in hindsight we evolved from small, dog-sized cynodonts that had to be nimble and quick to survive, hence the adaptive amplification of the fight or flight mechanism.

The environment gave rise to endothermy

All of the above has been incorporated into a "Central Theory of Biology" for the evolution of warm-bloodedness, or endothermy/homeothermy (Torday, 2015b). Briefly (see Fig. 3), the adrenalin that alleviated the lung alveolar constraint on oxygenation also increased the release of fatty acids from fat stores (Carey, 1998). Fatty acids are the optimal substrate for metabolic production of heat, so the increased metabolic activity would have raised body temperature,

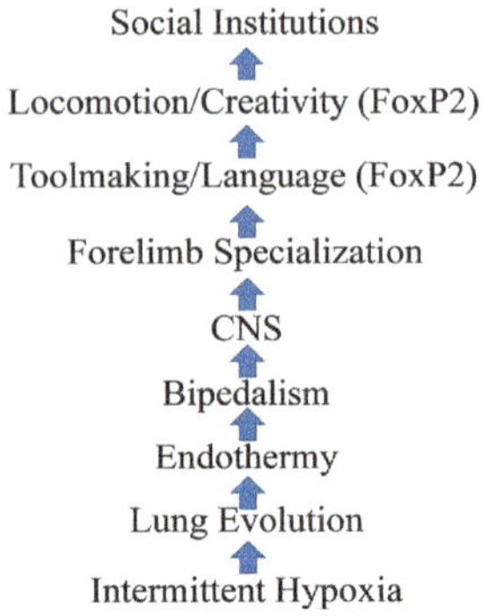

Fig. 3 Central theory of biology. Starting with intermittent hypoxia, the evolution of land vertebrates advanced due to endothermy, allowing for bipedalism, freeing the forelimbs for toolmaking, including language and imagination codified as the gene FoxP2. The development of written language facilitated the rise of human social institutions, allowing for the transfer of knowledge from generation to generation.

ultimately evolving genetic control as the thermoregulatory action of oxytocin (Kasahara et al., 2013), a neuroendocrine hormone produced by the posterior pituitary gland (Leng et al., 2015).

The ability to maintain body temperature independently of the environment fostered the transition from cold- to warm-bloodedness due to more efficient metabolism since multiple enzyme isoforms are needed for any given metabolic step in cold-blooded organisms in order to optimize the metabolic activity at different ambient environmental temperatures, whereas warm-blooded organisms only require one form of any given metabolic enzyme.

Increased metabolic efficiency facilitated bipedalism because it takes more energy to walk on two legs than on four (Rodman and McHenry, 1980), resulting in the freeing of the forelimbs for specialized adaptations such as flight in the case of birds and toolmaking in the case of humans. That placed additional selection pressure on the central nervous system to integrate and control such complicated physiologic properties. The upshot of that selection pressure was language as a form of toolmaking, particularly the written word, which codified the myths we use to cope with our ambiguous origin referred to above. And because such hard copies of our narratives can be passed from generation to generation, they become part and parcel of our institutionalized belief systems.

Only great apes have a structure on the underside of the cerebrum called the Area of Broca that controls both motor control and language (Palomero-Gallagher and Zilles, 2019). In humans, the coalescing of toolmaking, locomotion, and language is under the common control of FoxP2 (Xu et al., 2018), underscoring the positive selection for this constellation of traits in humans. And beyond that, FoxP2 gives rise to human creativity (Lieberman, 2009). It appears as though the combination of language and bipedal locomotion fosters our curiosity, giving rise to imagination.

The progressive increase in entropy in the environment due to the Second Law of Thermodynamics placed ever-greater selection pressure on biology to amplify the First Principles of Physiology (Torday and Rehan, 2009) that initially permitted negative entropy within the cell. That, in turn, would have been dependent on the lipid facilitation of oxygenation, initiated by the insertion of cholesterol into the cell membrane (Bastiaanse et al., 1997), followed by the evolution of peroxisome-derived lipids to protect against the rising oxygen within the cell (De Duve, 1969), lipids as substrate for steroid hormones (Hume and Boyd, 1978), the endocrine system (Kalra and Priya, 2018), and physiologic evolution (Torday and Rehan, 2012, 2017).

The recognition of lung surfactant evolution as a series of preadaptations (Torday and Rehan, 2009) provided deep insights into the fundamental interrelationship between lipids and oxygen uptake, from the composition of lung surfactant lipids (Orgeig et al., 2015) all the way back to the unicellular state based on the biosynthesis of cholesterol (Torday, 2013), the most primitive of lung surfactants (Orgeig and Daniels, 2001); Konrad Bloch hypothesized that cholesterol was a "molecular fossil" since it takes 11 atoms of oxygen to produce one molecule of cholesterol (Bloch, 1992). Therefore, there had to have been enough oxygen in the atmosphere to do so, linking oxygen and cholesterol together mechanistically in space-time.

Overall, that process, in turn, gave insight into the evolution of many other physiologic traits, particularly those facilitated by PTHrP, namely the lung, kidney, skin, brain, and skeleton (Guise and Mundy, 1996). Such insights led to a focus for vertebrate evolution on the water-land transition (Romer, 1949), during which the PTHrP Receptor gene duplicated (Pinheiro et al., 2012), or essentially amplified due to its property as a receptor.

Tiktaalik provided scientific evidence for the fossilized remains of the transition from fish to tetrapods (Daeschler et al., 2006), but there are no fossil data for the modifications of the internal organs that occurred during that transition in adaptation to land since such soft tissue structures would not have been preserved. However, there are extensive data for the cellular-molecular development of the organs that were essential for land adaptation, such as the lung and kidney. When such developmental mechanisms are superimposed on phylogenetic changes, they reflect the underlying cellular-molecular changes that occurred over the course of evolution (Torday and Rehan, 2007). Such hypotheses have been corroborated by gene deletions and overexpressions that are consistent with such evolutionary changes.

The Periodic Table and Evolutionary Biology are diachronic vectors of the Big Bang

The use of PTHrP signaling diachronically across space-time is homologous (of the same origin) with Mendeleev's use of chemical reactions to construct his Periodic Table of Elements. More importantly, receptor signaling through such "second messengers" as cyclic adenosine monophosphate and inositol phosphates (Rubin et al., 1994) gave even deeper insight into other such cell-cell signaling mechanisms occurring in tandem in other tissues and organs (Torday and Rehan, 2012). The magnitude and direction of these signaling pathways are vectors for biologic change, referring all the way back to the unicellular state (Torday and Miller, 2016), which dominated life on Earth for the first 3.5 billion years (Mojzsis et al., 1996). That pattern is comparable to the way in which chemical reactions aided Mendeleev's assembly of the Periodic Table of Elements. Importantly, both the chemical and biologic reactions are diachronic mechanisms that cut across space-time. The vectoral magnitude and direction of those chemical and biologic reactions ultimately reference the Singularity (Torday, 2018a), from which they emerged as asymmetries produced by the Big Bang. Admittedly, there is a 10-billion-year gap between the Big Bang and the origin of life, but the case has been made for the homology between the atom and the cell (Torday and Miller Jr., 2016b), providing the structural-functional bridge for that gap.

For example, this may be why William of Ockham declared that the simplest answer is probably correct (Spade, 1999). Those observations that are most consistent with the vector of the Big Bang represent the shortest distance between two points, namely the Singularity and physical, chemical, and biologic reactions.

Mathematical expression of such vectors would lead to a way of "calculating" the value of any given property of Nature. Peter Rowlands, a physicist at the University of Liverpool, has provided just such a mathematical system (Rowlands, 2015), arguing that there are fundamental principles of physics that underlie all we experience. In his reduction, all of reality can be expressed in terms of "zeros and ones." We conventionally focus on the "ones" as our material reality, but it is actually zero, which Rowlands refers to as an attractor, or organizer, that is the key to understanding the fundament of physics. That is because material things are merely epiphenomena of the energy produced by the Big Bang, matter being a by-product due to Newton's Third Law of Motion—every action having an equal and opposite reaction. Absent that "recoil," the Cosmos would only be composed of energy. The same holds true for biology; witness the morphologic changes that occur during embryology, beginning with the fertilization of the egg, characterized by a huge flash of energy due to the "zinc spark" (Zhang et al., 2016). The subsequent cell-cell interactions that mediate embryologic growth and differentiation are mediated by growth factors triggering high-energy phosphates. Focusing on the flow of energy exchanges rather than on the cellular changes provides a much simpler perspective on life (Torday, 2016) than Darwin's "tangled bank," particularly when it is considered that the developmental cell-cell interactions culminate in physiologic homeostasis as the energetic balance for the rest of the life cycle, the organism serving to collect epigenetic data from its environment (Torday and Miller Jr., 2016a).

When thought of in those terms, focusing on the cell as the basis for biologic evolution amounts to the same thing because the negative entropy within the cell is "zero" relative to the positive entropy outside of the cell (Schrodinger, 1944). Negative numbers are a mathematical construct that is not relevant to biology. Therefore, the negative entropy within the cell is virtually zero, but no less. Therefore, both the animate and the inanimate can be reduced to the same set of fundamental principles. Indeed, this way of thinking about Cosmology is synonymous with Alfred North Whitehead's Process Philosophy (Whitehead, 1929). He thought that everything in the Cosmos was energy and that matter occasionally appears coincidentally, but is transient.

Whitehead thought that our focus on the material at the expense of the nonmaterial was misguided because it did not engender relationships. In his book "Science and the Modern World," Whitehead (1925) expressed the idea that matter is "senseless, valueless, purposeless" because it is merely a transient by-product of the underlying energetic forces of the Cosmos. He called this perspective "scientific materialism." Whitehead found fault with the irreducible nature of matter because it masks the importance of change since nothing ever stays the same. Whitehead placed the emphasis of reality on change, and that "all things flow." This concept was first voiced by Heraclitus, who said that "No man ever steps in the same river twice, for it is not the same river and he is not the same man" (Kahn, 1979).

Whitehead thought that viewing objects as separate and distinct from all other objects was a systematic error because each object is an inert mass that is only superficially related to other things. The idea that the material is the primary state of being leads people to conclude that objects are all separated by time and space and are not related to anything.

Conversely, Whitehead thought that relationships were the primary state of being. Whitehead describes any entity being nothing more or less than the aggregate of its relationship to other entities, as the synthesis of and reaction to the world around it. Relationships are not secondary to what a thing is, they *are* what the thing is. Robert Rosen held a similarly strong opinion about relationships (Rosen, 1958).

Whitehead's concept of reality is consistent with Mendeleev's way of assembling the Periodic Table (Scerri, 2019). He saw Elements in the context of their reactivity with other elements rather than as inert matter, with certain superficial characteristics like those described by Alchemists. In that sense, it was the Alchemist who determined how to make gold out of dross, not the innate characteristics of the Elements that enabled them to interrelate with one another.

Similarly, by viewing organisms through the lens of their abilities to recapitulate themselves developmentally, the processes by which they fundamentally interrelate with one another come into view. That is particularly true when development is seen in the larger context of phylogeny, offering the retrograde retracing of the process of evolution back to the unicellular state as a series of preadaptations (Torday and Rehan, 2012). Such interactions between the organism and its environment over the course of evolution subordinate the material in favor of the process.

Therefore, now it can be stated unequivocally that both the Periodic Table of Elements and Evolutionary Biology are expressions of the process of change rather than descriptions of the superficial appearance of material objects—elements and phenotypes. Whitehead was thus correct in his perspective, but he had no experimental evidence to scientifically substantiate what he was espousing. Now, armed with evidence for this comprehensive view of reality, it behooves us to practice Whitehead Process Philosophy.

Information Theory meets Informatics

Information Theory is the epitome of Whitehead's focus on process since it represents the mechanism for forming, storing, and sharing information (Shannon, 1948). However, because it subordinates the transmission of the information to the materialism of the information, it is misguided. The consequence of that is the discipline of Informatics (Arndt, 2004), which tends to be conflated with knowledge. This is syllogistic reasoning that has undermined human thought and action. The prevailing concept in Informatics is that if you have not solved the problem, you just need more data. That may be true at NASA, where Informatics was developed to keep track of the parts for assembling the space shuttle, but does not apply to biologic systems.

In the case of the shuttle, the data are a closed set, whereas biology is an open set, the whole being greater than the sum of its parts.

For example, hormones do not carry information, they merely facilitate communication between cells, triggering the energy bursts of "second messengers," which are high-energy phosphate compounds that bind to DNA elements in the nucleus and affect embryologic growth and differentiation. It is only once the hormone binds to its receptor that information within the cell is made available. Therefore, the information lies within the target cell itself, not in the hormone.

Similarly, in the article about "Phenotype as Agent" (Torday and Miller Jr., 2016a), it is argued that we misconstrue what phenotype is based on the description rather than the mechanism of evolution. Phenotype is the agency of the Explicate Order for detecting environmental change, whereas convention says that it is the material composition of the organism. In this frame, perhaps our physiology is the product of the Implicate Order, acting as the reference point for the phenotypic Explicate Order. The residue of the Implicate Order remains in our physiology in various forms, such as the mesoderm, which is introduced between the endoderm and ectoderm as the third germ layer during gastrulation. Lewis Wolpert said it was "truly the most important time in your life," because the mesoderm confers the plasticity to adapt. And then, there's the advent of deuterostomes—developing from posterior to anterior, in the opposite direction to the force of gravity. That is homologous with the effect of bipedalism, freeing the forelimbs to specialize, again placing positive selection pressure anteriorly for further evolution of the head and brain (Torday, 2015b). That process is mirrored and reinforced by the vagus, the major nerve of the autonomic nervous system, which emanates from the adrenals, to the gut, to the heart, terminating in the head (Porges, 1995).

Truth be told

It is because of the relationship between materialism and process that we have been able to advance as a species among species. However, as David Bohm points out, we continue to exist within the Explicate

Order, made subjective by our evolved senses, whereas there is a true reality just beyond our reach, referred to as the "Implicate Order" (Bohm, 1980). It is because of the deceptive reality we have formulated (Trivers, 2011) that we must periodically rise up, and then fall, ultimately succumbing to the Laws of Nature. Whether it is Climate Change or economics, we are vulnerable to our failure to comply fully with the prevailing forces of Nature. We come ever-closer to those Laws as we evolve and endogenize the environment over the course of evolution, but in the current environment of the Anthropocene (Zalasiewicz et al., 2010), our narcissistic tendencies are out-stripping our natural arc for lack of insight regarding process over materialism (Whitehead, 1929). Prior to the use of fossil fuel to drive our machines, there was an intimate, demonstrable interrelationship between the energy of the Earth and that of life (Torday and Rehan, 2011). Subsequently, the introduction of man-made thermal energy into the environment caused a dissociation of Man from our evolutionary trajectory. Consequently, once we begin engineering our heredity using CRISPR (Zhang et al., 2014), we will deviate from our naturally evolved trajectory from the Singularity, perhaps evolving as "silicon-based life forms" instead.

There is only space, there is no time

If, as Rowlands has concluded, the only directly measurable dimension is space (Rowlands, 2015), Einstein and Feynman (Feynman, 2011) having cast doubt on the existence of time, how do we explain the time-based understanding of development and phylogeny? It would have to be assumed that the latter are exclusively space-filling properties of biology. In this vein, we have learned that epigenetic inheritance directly from the environment affects evolution, and it has been proposed that it fosters "running in place" to maintain homeostasis in consilience with the First Principles of Physiology (Torday and Rehan, 2009), which would subsume a spatially non-temporal way of thinking about biology. This question is reminiscent of the well-documented debate between Einstein and Bergson in 1922, Einstein insisting that time is an artifact of biology, Bergson countering that time is critical for understanding any and all of biology and psychology (Canales, 2016).

If, as has been proposed (Torday, 2019), the Singularity is the prototype for biology, and evolution is the process for remaining faithful to it, striving to emulate the Singularity, then time would drop out of the "equation," leaving space as a point source at its minimum (Torday and Miller Jr., 2016b) and the Cosmos as its maximum.

A novel prediction of Consciousness as the Singularity

The premise for the idea that Consciousness is the expression of the Singularity is that there is an intersection of the Singularity with physiology, the latter as the effective endogenization of the cosmologic environment (Torday, 2020), at least those features that have posed existential threats to the organism. That is to say, when the Singularity was disrupted by the Big Bang (see Fig. 4), the information therein was fragmented, but still ultimately had to conform with the Laws of Nature. When life began on Earth some 4.5 billion years ago, it too had to conform with the Laws of Nature. It did so by endogenizing the environment and making it useful by compartmentalizing it as what we refer to as "physiology" (Margulis, 1993). In the aggregate, our physiology ascribes to the Singularity as its origin (Torday, 2019), and the way in which physiology functions to maintain homeostasis

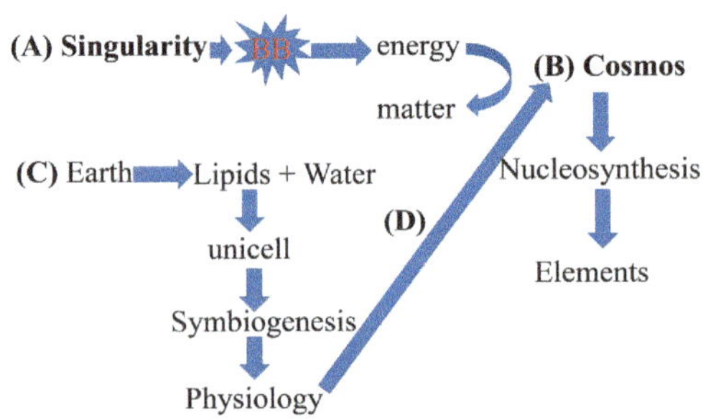

Fig. 4 The origin of Consciousness. The Big Bang (A) gave rise to energy and matter. (B) Nucleosynthesis, the formation of light by stars, generated the Elements in sequence from the lightest to the heaviest, providing a "logic." (C) Cells were formed from lipids immersed in water; when posed with existential threats, cells assimilated them through Symbiogenesis. Such factors formed physiologic systems, (D) but must comply with the same Laws of Nature as the Cosmos, forming the connection between life and nonlife.

is based on the same set of principles (Torday, 2015c). In other words, our interoceptive sense of self is founded on the Singularity as the origin of consciousness, actualized by the physiologic principles that have evolved from the Cosmos (Torday, 2018b). In turn, our agency as conscious beings gains purchase to ongoing variations in our ever-changing environment.

Holistically, our physiology maps onto the Singularity since it is the product of endogenized physical factors in the environment that have been compartmentalized and made functionally useful (Torday, 2019). Physiology functions to maintain homeostasis based on the same set of natural laws that the atom ascribes to (Torday, 2015c). When lipids spontaneously formed micelles, distinguishing the inside of the cell from the outside environment, fueled by negative entropy (Schrodinger, 1944), life began, initiating the Explicate Order in the process (Bohm, 1980). Consequently, our interoceptive sense of "inside" connects to the Singularity as consciousness, actualized by the physiologic principles that have evolved from the Cosmos (Torday, 2018b).

It is the principle of homeostasis that integrates all of these properties—when the Big Bang occurred some 13.8 billion years ago, there was an "equal and opposite reaction" to it based on Newton's Third Law of Motion. That reaction is the preadaptation that we refer to as homeostasis, acting to entrain inanimate balanced physical, chemical, and biologic reactions alike (Torday, 2015c). Without homeostasis, there would be no matter, there would only be energy. Alfred North Whitehead's "Process Philosophy" states that all is energy; matter is a transition between energy states (Whitehead, 1929).

The vertical integration of gravity, chemistry, and biology as Consciousness

As mentioned above, PTHrP integrated numerous physiologic properties in vertebrates due to evolutionary selection pressure during the water-land transition. This is attributable to its mechanotransduced effects on the skeleton and lung. Such vertically integrated physiologic mechanisms offer deep insights into the fundamental interrelationships between physics, chemistry, and biology. For example, when lung or bone cells are exposed to microgravity, PTHrP messenger RNA decreases rapidly, interrupting the cell-cell communication mechanisms it mediates for breathing (lung), salt/water balance (kidney), and bone calcification, for example (Torday, 2003). In addition to being a mechanotransducer for cell-cell interactions, PTHrP is a calcium regulatory paracrine factor, acting synergistically with its effect on cell-cell communication to ensure that calcium levels in the alveolar hypophase are appropriate for optimal lung surfactant surface tension-lowering function, and for calcification of the skeleton in response to PTHrP signaling in bone (Torday, 2003). Even more fundamentally, when yeast are exposed to microgravity, they lose their ability to generate a calcium flux or reproduce, leaving them "comatose" (Purevdorj-Gage et al., 2006). Molecularly, mechanical forces like gravity affect the cell via the Target of Rapamycin (TOR) gene, which interacts with the cytoskeleton (Diz-Muñoz et al., 2016). In turn, matrix proteins in the extracellular space interconnect with cytoplasmic signaling mechanisms, orchestrating the myriad chemical reactions that govern cell physiology on a moment-to-moment basis. Thus, the gravitational force produced by the Big Bang references the Singularity as the physiologic unity being emulated by the unicell, endogenized over the course of evolution, ultimately generating Consciousness as the holism of being (Torday and Miller Jr., 2018).

For example, it has been shown that gravity is necessary for life since differentiated lung and bone cells lose their phenotypic identity in zero gravity (Torday, 2003). Similarly, Purevdorj-Gage et al. (2006) have shown that yeast, which are simple eukaryotes, exposed to zero gravity cannot conduct a calcium flux because they cannot polarize, and cannot reproduce, rendering them nonfunctional. Even the purely chemical micelle cannot form normally in the absence of gravity (Claassen and Spooner, 1996). That is because the fundament of Symbiogenesis is Quantum Entanglement, which entrained particles within micelles at the beginning of life (Torday, 2021), fueled by gravity impinging on the curved surface of the micelle (Misner et al., 1973), the particles synchronizing themselves nonlocally with events in the Cosmos. This continuum from the origin of life to the Cosmos is where consciousness evolved from (Torday, 2018b).

Biology and chemistry as vectoral fractals of the Big Bang

As mentioned at the outset, the genius of Mendeleev's Periodic Table of Elements was not only in his use of atomic weight as the organizing principle, but in further fine-tuning it based on the chemical reaction products that further characterized each element (Scerri, 2019). That empiric approach offered a much more dynamic way of calibrating the Elements diachronically across space-time rather than merely arranging them based on their physical characteristics synchronically.

The same holds true for using the cell-cell interactions that generate form and function developmentally, applied to phylogeny to understand evolution, which is also a diachronic perspective. The value added by this perspective is in being able to eliminate time from the analysis, leaving only the dimension of space to be considered; and space can be reduced to a point-source without any dimensions, as pure energy, consistent with Whitehead's "Process Theory." In contrast to that approach, forming cladograms merely describes the physical progression of evolution without any underlying understanding of how or why it occurred at the cellular-molecular level.

Moreover, the use of such empiric data reflects the changes in the chemical and biologic reactants to their products. When shown graphically, those reactions are vectors for the magnitude and direction of change. It is hypothesized that those vectors are fractals of the Ur-vector formed by the Big Bang. And the more proximate the chemical and evolutionary biologic vectors are to the vector formed by the Big Bang, the more fundamental they are. Yet like Zeno's Paradox, we can never be evolutionarily congruent with the ultimate vector of the Big Bang because we ironically must evolve in response to vectoral changes in the environment or become extinct.

This is why William of Occam's Razor is indicative of the right solution, the shortest distance between two points being the shortest, simplest path.

Terminal Addition and Phantom Limb as "Proof of Principle" for Physiology as our Implicate Order

The phenomenon of Terminal Addition, or the appearance of a newly acquired trait at the end of a series of evolved traits, is well recognized in biology, yet how and why it occurs has not been explained. It stems from the fact that embryologic development is achieved by cell-cell communications mediated by soluble growth factors signaling to their cognate receptors, stimulating the production of high-energy phosphates as "second messengers" for growth or differentiation. The most efficient way to introduce such a new trait is by adding it on to the end terminus because inserting in the middle or at the beginning would disrupt the predecessors, requiring them to rewire, which is highly inefficient, putting the organism at risk of extinction, whereas adding such new traits to the end is relatively easy. In so doing, it provides a "logic" founded on the assimilation of environmental factors, aligning the internal environment of the cell with the external environment of the environment, providing a natural homology between the cell and the Cosmos.

A concomitant of Terminal Addition is the phenomenon of "Phantom Limb Syndrome," an amputee having an itch in the big toe of a severed leg. At first glance, this sense seems highly inefficient. However, when seen in light ofTerminal Addition, loss of such a phantom sensation would result in loss of sensation in the traits upstream from the amputated limb too, causing their atrophy. So for the sake of conservation of structure and function, an imagined input from the missing limb is maintained in order to preserve the integrity of the organism in the vein the phenotype as agent.

Conclusions

In an earlier publication, it had been suggested that regressing the data for cellular-molecular lung evolution would approximate the origin of life at the intersection of its Cartesian coordinates

(Torday and Rehan, 2007) by reverse-engineering physiologic traits like the lung, tracing the cell-cell communications from the alveolus back to the swim bladder, and further back to the cell membrane of unicellular eukaryotes based on lipids facilitating oxygen uptake (see Fig. 1). Using that approach allows a "back-calculation" as far back as Quantum Entanglement and nonlocalization as preadaptations for Symbiogenesis.

The present claim for that prediction is also based on cellular evolution, but now as serial preadaptations, specifically identifying the Singularity as the ultimate preadaptation of the unicellular state (Torday, 2019). As a "point source" (Torday and Miller Jr, 2017), the Singularity would occupy no space, merely being a locale of energy essentially equaling zero space. Rowlands has similarly suggested that in math, zero is an attractor, acting as an organizing principle (Rowlands, 2015).

Such metaphysical ideas might help in identifying the actual nature of Consciousness, which has remained intractable for thousands of years, beginning with the Ancient Greek Philosophers, right up to today, when philosophers like David Chalmers (Chalmers, 1996) and Andy Clark (Clark and Chalmers, 1998) pose "hard" questions about qualia. In the current context, it is proposed that Consciousness lies at the intersection of Cosmology and Physiology, the product of which is what we think of as being conscious, or mind.

It is offered that chemical and physical properties are vectoral fractals of the Big Bang, which could be tested by mathematically modeling key reactions for each.

References

Arndt, C., 2004. Information Measures, Information and its Description in Science and Engineering. Springer, New York.

Bastiaanse, E.M., Höld, K.M., Van der Laarse, A., 1997. The effect of membrane cholesterol content on ion transport processes in plasma membranes. Cardiovasc. Res. 33, 272–283.

Berner, R.A., 1999. Atmospheric oxygen over Phanerozoic time. Proc. Natl. Acad. Sci. U. S. A. 96, 10955–10957.

Bloch, K., 1992. Sterol molecule: structure, biosynthesis, and function. Steroids 57, 378–383.

Bohm, D., 1980. Wholeness and the Implicate Order. Routledge, London.

Canales, J., 2016. The Physicist and the Philosopher. Princeton University Press, Princeton.

Carey, G.B., 1998. Mechanisms regulating adipocyte lipolysis. Adv. Exp. Med. Biol. 441, 157–170.

Chalmers, D., 1996. The Conscious Mind. Oxford University Press, Oxford.
Claassen, D.E., Spooner, B.S., 1996. Liposome formation in microgravity. Adv. Space Res. 17, 151–160.
Clark, A., Chalmers, D.J., 1998. The extended mind. Analysis 58, 7–19.
Daeschler, E.B., Shubin, N.H., Jenkins Jr., F.A., 2006. A Devonian tetrapod-like fish and the evolution of the tetrapod body plan. Nature 440, 757–763.
Darwin, C., 1859. On the Origin of Species. John Murray, London.
De Duve, C., 1969. Evolution of the peroxisome. Ann. N.Y. Acad. Sci. 168, 369–381.
Di Corpo, U., Vannini, A., 2015. Syntropy. ICRL Press, Princeton.
Diamond, A.G., Gonterman, R.M., Anderson, A.L., Menon, K., Offutt, C.D., Weaver, C.H., Philbrick, W.M., Foley, J., 2006. Parathyroid hormone hormone-related protein and the PTH receptor regulate angiogenesis of the skin. J. Invest. Dermatol. 126, 2127–2134.
Diz-Muñoz, A., Thurley, K., Chintamen, S., Altschuler, S.J., Wu, L.F., Fletcher, D.A., Weiner, O.D., 2016. Membrane tension acts through PLD2 and mTORC2 to limit actin network assembly during neutrophil migration. PLoS Biol. 14, e1002474.
Dodelson, S., Schmidt, F., 2020. Modern Cosmology. Academic Press, San Diego.
Einstein, A., 1961. Relativity. The Special and General Theory. Crown Publishing, New York.
Feynman, R., 2011. The Feynman Lectures. Basic Books, New York.
Guex, J., Torday, J.S., Miller Jr., W.B., 2020. Morphogenesis, Environmental Stress and Reverse Evolution. Springer, Switzerland.
Guise, T.A., Mundy, G.R., 1996. Physiological and pathological roles of parathyroid hormone-related peptide. Curr. Opin. Nephrol. Hypertens. 5, 307–315.
Hume, R., Boyd, G.S., 1978. Cholesterol metabolism and steroid-hormone production. Biochem. Soc. Trans. 6, 893–898.
Kahn, C., 1979. The Art and Thought of Heraclitus: Fragments with Translation and Commentary. Cambridge University Press, London.
Kalra, S., Priya, G., 2018. Lipocrinology—the relationship between lipids and endocrine function. Drugs Context 7, 212514.
Kasahara, Y., Sato, K., Takayanagi, Y., Mizukami, H., Ozawa, K., Hidema, S., So, K.H., Kawada, T., Inoue, N., Ikeda, I., Roh, S.G., Itoi, K., Nishimori, K., 2013. Oxytocin receptor in the hypothalamus is sufficient to rescue normal thermoregulatory function in male oxytocin receptor knockout mice. Endocrinology 154, 4305–4315.
Kuhn, T., 1957. The Copernican Revolution. Harvard University Press, Cambridge.
Lawson, E.E., Brown, E.R., Torday, J.S., Madansky, D.L., Taeusch Jr., H.W., 1978. The effect of epinephrine on tracheal fluid flow and surfactant efflux in fetal sheep. Am. Rev. Respir. Dis. 118, 1023–1026.
Leng, G., Pineda, R., Sabatier, N., Ludwig, M., 2015. 60 years of neuroendocrinology: the posterior pituitary, from Geoffrey Harris to our present understanding. J. Endocrinol. 226, T173–T185.
Libet, B., Gleason, C.A., Wright, E.W., Pearl, D.K., 1983. Time of conscious intention to act in relation to onset of cerebral activity (readiness-potential). The unconscious initiation of a freely voluntary act. Brain 106, 623–642.
Lieberman, P., 2009. FOXP2 and human cognition. Cell 137, 800–802.
Mamillapalli, R., Wysolmerski, J., 2010. The calcium-sensing receptor couples to Galpha(s) and regulates PTHrP and ACTH secretion in pituitary cells. J. Endocrinol. 204, 287–297.
Margulis, L., 1993. Origins of species: acquired genomes and individuality. Biosystems 31, 121–125.
Mazzocchi, G., Aragona, F., Malendowicz, L.K., Nussdorfer, G.G., 2001. PTH and PTH-related peptide enhance steroid secretion from human adrenocortical cells. Am. J. Physiol. Endocrinol. Metab. 280, E209–E213.
Mehra, J., Rechenberg, H., 1982. The Historical Development of Quantum Theory, Vol. 1: The Quantum Theory of Planck, Einstein, Bohr and Sommerfeld. Its Foundation and the Rise of its Difficulties (1900–1925). Springer-Verlag, New York.

Misner, C.W., Thorne, K.S., Wheeler, J.A., 1973. Gravitation. W.H. Freeman, San Francisco.
Mojzsis, S.J., Arrhenius, G., McKeegan, H.T.M., Nutman, A.P., Friend, C.R., 1996. Evidence for life on earth before 3,800 million years ago. Nature 384, 55–59.
Orgeig, S., Daniels, C.B., 2001. The roles of cholesterol in pulmonary surfactant: insights from comparative and evolutionary studies. Comp. Biochem. Physiol. A Mol. Integr. Physiol. 129, 75–89.
Orgeig, S., Morrison, J.L., Daniels, C.B., 2015. Evolution, development, and function of the pulmonary surfactant system in normal and perturbed environments. Compr. Physiol. 6, 363–422.
Palomero-Gallagher, N., Zilles, K., 2019. Differences in cytoarchitecture of Broca's region between human, ape and macaque brains. Cortex 118, 132–153.
Pinheiro, P.L., Cardoso, J.C., Power, D.M., Canário, A.V., 2012. Functional characterization and evolution of PTH/PTHrP receptors: insights from the chicken. BMC Evol. Biol. 12, 110.
Popper, K., 1959. The Logic of Scientific Discovery. Routledge, London.
Porges, S.W., 1995. Orienting in a defensive world: mammalian modifications of our evolutionary heritage. A Polyvagal Theory. Psychophysiology 32, 301–318.
Purevdorj-Gage, B., Sheehan, K.B., Hyman, L.E., 2006. Effects of low-shear modeled microgravity on cell function, gene expression, and phenotype in Saccharomyces cerevisiae. Appl. Environ. Microbiol. 72, 4569–4575.
Rodman, P.S., McHenry, H.M., 1980. Bioenergetics and the origin of hominid bipedalism. Am. J. Phys. Anthropol. 52, 103–106.
Romer, A.S., 1949. The Vertebrate Story. University of Chicago Press, Chicago.
Rosen, R., 1958. A relational theory of biological systems. Bull. Math. Biophys. 20, 245–260.
Rowlands, P., 2015. The Foundations of Physical Law. World Scientific Publishing, Singapore.
Rubin, L.P., Kifor, O., Hua, J., Brown, E.M., Torday, J.S., 1994. Parathyroid hormone (PTH) and PTH-related protein stimulate surfactant phospholipid synthesis in rat fetal lung, apparently by a mesenchymal-epithelial mechanism. Biochim. Biophys. Acta 1223, 91–100.
Rubin, L.P., Kovacs, C.S., De Paepe, M.E., Tsai, S.W., Torday, J.S., Kronenberg, H.M., 2004. Arrested pulmonary alveolar cytodifferentiation and defective surfactant synthesis in mice missing the gene for parathyroid hormone-related protein. Dev. Dyn. 230, 278–289.
Scerri, E.R., 2019. The Periodic Table: Its Story and its Significance. Oxford University Press, Oxford.
Schrodinger, E., 1944. What is Life? Cambridge University Press, Cambridge.
Shannon, C.E., 1948. A mathematical theory of communication. Bell Syst. Tech. J. 27, 379–423. 623–656.
Sharara-Chami, R.I., Joachim, M., Pacak, K., Majzoub, J.A., 2010. Glucocorticoid treatment—effect on adrenal medullary catecholamine production. Shock 33, 213–217.
Smolin, L., 1999. The Life of the Cosmos. Oxford University Press, Oxford.
Spade, P.V., 1999. The Cambridge Companion to Ockham. Cambridge University Press, Cambridge.
Torday, J.S., 2003. Parathyroid hormone-related protein is a gravisensor in lung and bone cell biology. Adv. Space Res. 32, 1569–1576.
Torday, J.S., 2013. Evolution and cell physiology. 1. Cell signaling is all of biology. Am. J. Physiol. Cell Physiol. 305, C682–C689.
Torday, J.S., 2015a. The cell as the mechanistic basis for evolution. Wiley Interdiscip. Rev. Syst. Biol. Med. 7, 275–284.
Torday, J.S., 2015b. A central theory of biology. Med. Hypotheses 85, 49–57.
Torday, J.S., 2015c. Homeostasis as the mechanism of evolution. Biology (Basel) 4, 573–590.
Torday, J.S., 2016. Life is simple-biologic complexity is an epiphenomenon. Biology (Basel) 5, 17.

Torday, J.S., 2018a. Quantum mechanics predicts evolutionary biology. Prog. Biophys. Mol. Biol. 135, 11–15. 2018.
Torday, J.S., 2018b. From cholesterol to consciousness. Prog. Biophys. Mol. Biol. 132, 52–56.
Torday, J.S., 2019. The singularity of nature. Prog. Biophys. Mol. Biol. 142, 23–31.
Torday, J.S., 2020. Consciousness, redux. Med. Hypotheses 140, 109674.
Torday, J.S., 2021. Life is a mobius strip. Prog. Biophys. Mol. Biol. 167, 41–45.
Torday, J.S., 2022. Hormones and Reality. Elsevier, Amsterdam.
Torday, J.S., Miller Jr., W.B., 2017. The resolution of ambiguity as the basis for life: a cellular bridge between Western reductionism and Eastern holism. Prog. Biophys. Mol. Biol. 131, 288–297.
Torday, J.S., Miller Jr., W.B., 2016a. Phenotype as agent for epigenetic inheritance. Biology (Basel) 5, 30.
Torday, J.S., Miller Jr., W.B., 2016b. The unicellular state as a point source in a quantum biological system. Biology (Basel) 5, 25.
Torday, J.S., Miller Jr., W.B., 2018. Unitary physiology. Compr. Physiol. 8, 761–771.
Torday, J.S., Rehan, V.K., 2007. The evolutionary continuum from lung development to homeostasis and repair. Am. J. Physiol. Lung Cell. Mol. Physiol. 292, L608–L611.
Torday, J.S., Rehan, V.K., 2009. The evolution of cell communication: the road not taken. Cell Commun. Insights 2, 17–25.
Torday, J.S., Rehan, V.K., 2011. A cell-molecular approach predicts vertebrate evolution. Mol. Biol. Evol. 28, 2973–2981.
Torday, J.S., Rehan, V.K., 2012. Evolutionary Biology, Cell-Cell Communication and Complex Disease. Wiley, Hoboken.
Torday, J.S., Rehan, V.K., 2017. Evolution, the Logic of Biology. Wiley, Hoboken.
Trivers, R., 2011. The Folly of Fools. Basic Books, New York.
Vermeij, G.J., 2016. Gigantism and its implications for the history of life. PLoS One 11, e0146092.
Whitehead, A.N., 1925. Science and the Modern World. MacMillan, New York.
Whitehead, A.N., 1929. Process and Reality. Macmillan, New York.
Wurtman, R.J., 2002. Stress and the adrenocortical control of epinephrine synthesis. Metab. Clin. Exp. 51, 11–14.
Xu, S., Liu, P., Chen, Y., Chen, Y., Zhang, W., Zhao, H., Cao, Y., Wang, F., Jiang, N., Lin, S., Li, B., Zhang, Z., Wei, Z., Fan, Y., Jin, Y., He, L., Zhou, R., Dekker, J.D., Tucker, H.O., Fisher, S.E., Yao, Z., Liu, Q., Xia, X., Guo, X., 2018. Foxp2 regulates anatomical features that may be relevant for vocal behaviors and bipedal locomotion. Proc. Natl. Acad. Sci. U. S. A. 115, 8799–8804.
Young, T., 1804. The Bakerian lecture. Experiments and calculation relative to physical optics. Philos. Trans. R. Soc. Lond. 94, 1–16.
Zalasiewicz, J., Williams, M., Steffen, W., Crutzen, P., 2010. The new world of the Anthropocene. Environ. Sci. Technol. 44, 2228–2231.
Zhang, N., Duncan, F.E., Que, E.L., O'Halloran, T.V., Woodruff, T.K., 2016. The fertilization-induced zinc spark is a novel biomarker of mouse embryo quality and early development. Sci. Rep. 6, 22772.
Zhang, F., Wen, Y., Guo, X., 2014. CRISPR/Case9 for genome editing: progress, implications and challenges. Hum. Mol. Genet. 23, R40–R46.
Zheng, W., Wang, Z., Collins, J.E., Andrews, R.M., Stemple, D., Gong, Z., 2011. Comparative transcriptome analyses indicate molecular homology of zebrafish swimbladder and mammalian lung. PLoS One 6, e24019.

CHAPTER 13

The latticework of both the Cosmos and physiology

Introduction

The Greeks spoke of the "One," Einstein equated energy and matter, and you may wonder from time to time why there is never a bad color combination in Nature. Hypothetically, it is because there is a "Singularity of Nature" (Torday, 2019). But that begs the question as to how and why that occurred. There are hints in the science that offer an answer, if only we were to use our imagination, which Einstein said was far more important than knowledge. I have employed the concept of a latticework to identify the common motif between the Cosmos and our physiology as follows.

Nucleosynthesis—First 36 Elements, all relevant to physiology with the exception of iodine, 53, which became adapted by the thyroid

Stars generate light through the process of nucleosynthesis. In so doing, they produce the first 36 Elements of the Periodic Table, many of which have been exploited for physiologic function. The only exception is iodine, Element 53, which was used by the thyroid gland to synthesize thyroxin by combining it with tyrosine. The thyroid evolved from the endostyle of elasmobranchs.

The phylogenetic history of the endostyle and nonfollicular thyroid tissues of vertebrates and invertebrates has been reviewed by Eales. He underscored the importance of exogenous compounds and peripheral mechanisms in regulating thyroidal status and in influencing the course of the phylogenetic development of the vertebrate thyroid gland. Key to understanding thyroid-like function is the ability to utilize iodine, and there are many nonchordate invertebrates

Quantum Mechanics, Cell-Cell Signaling, and Evolution
https://doi.org/10.1016/B978-0-323-91297-6.00013-8

that have the ability to bind iodine. Eaten and/or absorbed free iodo compounds, perhaps even iodothyronines from plants and microorganisms, can be metabolized in the gut with the aid of resident bacteria. Absorption of iodo compounds has not been shown in the larval lamprey intestine, but this organ is the primary site for monodeiodination from thyroxin to triiodothyronine, as is also the case for ascidians. The endostyle is found only in marine invertebrates with a notochord, and in freshwater larval lampreys. Although this specialized region of the alimentary tract may have evolved following selection pressure favoring filter feeding, like many tissues of invertebrates it had iodine-binding capacity, a function that developed secondarily. The iodine-binding capacity of the endostyle in extant ascidians and larval lampreys is well established. The retention of the endostyle throughout the evolutionary history of lampreys reflects its importance in the filter-feeding apparatus and also suggests that it may have appeared when lampreys were in their pelagic marine phase.

The vertebrate follicular thyroid gland evolved following strong positive selection pressure for a gland that favored thyroid hormone and iodine storage during the period when ancient chordates moved from their iodine-rich marine environment to their iodine-poor freshwater habitat. In this interpretation, the presence of an endostyle in extant larval lampreys is a reflection of their ancient marine origins, and the adult follicular thyroid gland came after they moved toward freshwater. Since the follicular thyroid gland of adult lampreys appears only following transformation of the endostyle during metamorphosis, there was selection pressure created by the freshwater environment for metamorphosis in the ontogeny of lampreys. A consequence of this metamorphosis was transformation of the endostyle into a follicular thyroid gland.

This view of the ontogeny of the lamprey thyroid gland, in conjunction with the belief that the freshwater habitat of lampreys is a secondarily acquired niche, implies that metamorphosis might not have originated as a developmental strategy, but occurred when lampreys moved into freshwater during their evolutionary history. An alternative view would be that the presence of metamorphosis, and ultimately the follicular thyroid gland in the marine environment, was the primary reason why lampreys could inhabit freshwater. The presence of an endostyle in early lampreys was not a requisite for living in a marine (brackish) environment, since modern-day larvae with endostyles cannot tolerate even dilute seawater. Premetamorphic lampreys have lost the ability of the presumed (hypothetical) ancestral larval form to inhabit a marine environment of any type. It is only after metamorphosis that juveniles of some species can tolerate seawater.

Larval and reproductive intervals of the lamprey life cycle are confined to freshwater. The ability of juveniles of some species to osmoregulate in seawater could also be a character that was secondarily derived following the appearance of metamorphosis in the ontogeny of lampreys. Extant parasitic lampreys of the most ancient lineage (e.g., *Ichthyomyzon unicupis*) are confined to freshwater. An explanation of how marine osmoregulation may have been secondarily derived following the primary derivation of metamorphosis may be found in modern views of the phenomenon called "developmental integration."

Evolutionary vertical integration of the phylogeny and ontogeny of the thyroid

Mechanistically, the increased bacterial load due to the facilitation of feeding by the endostyle may have stimulated the cyclic AMP-dependent protein kinase A, or PKA pathway, since bacteria produce endotoxin, a potent PKA agonist. This cascade may have evolved into regulation of the thyroid by thyroid-stimulating hormone (TSH), since TSH acts on the thyroid via the cAMP-dependent PKA signaling pathway. This mechanism potentially generated novel structures such as the thyroid, lung, and pituitary, all of which are induced by the PKA-sensitive Nkx2.1/TTF-1 pathway. The brain-lung-thyroid syndrome (Carré et al., 2009), in which infants with Nkx2.1/TTF-1 mutations develop hypotonia, hypothyroidism, and respiratory distress syndrome, or surfactant deficiency disease, is further evidence for the coevolution of the lung, thyroid, and pituitary.

Developmentally, the thyroid evaginates from the foregut beginning on day 8.5 in the mouse, 1 day prior to the lung and pituitary emerging, suggesting that the thyroid may have been a molecular prototype of the lung during evolution, providing a testable and refutable hypothesis. The thyroid rendered molecular iodine in the environment bioavailable by binding it to threonine to generate thyroid hormone, whereas the lung made molecular oxygen bioavailable, first by inducing fat cell-like lipofibroblasts as cytoprotectants, which then stimulated surfactant production by producing leptin, placing increased selection pressure on the blood-gas barrier by making the alveoli more compliant. This, in turn, may have created further selection pressure for the metabolic system to utilize the rising oxygen level in the environment, placing further selection pressure on the alveoli, giving rise to the stretch-regulated surfactant system mediated by the Parathyroid Hormone-related Protein and leptin. Subsequent selection pressure on the cardiopulmonary system may have facilitated liver evolution, since the progressively increasing size of the heart may have induced precocious liver development, fostering increased glucose regulation. The brain serves as a glucose sink, and there is experimental evidence that increasing glucose during pregnancy increases the size of the developing brain (Saintonge and Côté, 1984). The further evolution of the brain, specifically the pituitary, would have served to further the evolution of complex physiologic systems.

Symbiogenesis, Terminal Addition, and cell-cell signaling

The phenomenon of Terminal Addition in biology is well recognized. Newly acquired evolutionary traits are tacked on to the end of a series of evolved traits, rather than in the middle or at the beginning. The reason for this is explained in "Terminal Addition in a Cellular World" (Torday and Miller Jr., 2018) as follows:

> *Recent advances in our understanding of evolutionary development permit a reframed appraisal of Terminal Addition as a continuous historical process of cellular-environmental complementarity. Within this frame of reference, evolutionary terminal additions can be identified as environmental induction of episodic adjustments to cell-cell signaling patterns that yield the cellular-molecular pathways that lead to differing developmental forms. Phenotypes derive, thereby, through cellular mutualistic/competitive niche constructions in reciprocating responsiveness to environmental stresses and epigenetic impacts. In such terms, Terminal Addition flows according to a logic of cellular needs confronting environmental challenges*

over space-time. A reconciliation of evolutionary development and Terminal Addition can be achieved through a combined focus on cell-cell signaling, molecular phylogenies and a broader understanding of epigenetic phenomena among eukaryotic organisms. When understood in this manner, Terminal Addition has an important role in evolutionary development, and chronic disease might be considered as a form of 'reverse evolution' of the self-same processes.

When Terminal Addition is seen in the context of Symbiogenesis (Sagan, 1967), the endogenization and assimilation of environmental factors, it is the "logic" of the Cosmos that is dictating both the inanimate and the animate.

Gravity is necessary for life and forms the basis for the lattice

And beyond the mechanism of Terminal Addition, the force of gravity has formed the basis for both planets (Wurm and Teiser, 2021) and cells (Torday, 2003; Purevdorj-Gage et al., 2006) alike. As one of the four strong forces produced by the Big Bang, gravity has been found to underlie the formation of planetary spheres and cellular spheres alike. It is the resistance to the gravitational force that likely produced life, given that in the virtual absence of gravity differentiated cells lose their phenotypes, and in the case of yeast, they cannot polarize or bud, i.e., reproduce.

Synchronicity as a product of latticework

It was Carl Jung who hypothesized what he called "synchronicities," or things that happened concurrently (Sacco, 2019) due to different causes. This may be a consequence of the interrelationship between the force of gravity impinging on the curvature of the protocell producing energy, maintaining Quantum Entanglement (QE) and nonlocalization of gravity. In other words, spiritualism and science both emerge from the same effect of gravity.

Phantom Limb, Terminal Addition, and latticework

Phantom Limb is a well-recognized phenomenon—the sense of feeling in a severed limb. It makes no intuitive sense to expend physical and psychic energy on a nonexistent limb, but thought of in the context of Terminal Addition (see above), the cell-cell signaling proximal to the severed limb must be maintained to avoid disuse atrophy. This mechanism is consistent with the latticework of our physiology.

Quantum Mechanics and latticework: Quantum Chromodynamics theory

Working back from existing complicated physiology by exploiting cell-cell communication leads universally to the unicellular stage as a function of Symbiogenesis. But that raised the question as to where Symbiogenesis emerged from as a reference to both the internal and environmental conditions. Recalling that Einstein's Field Theory dictates that when gravity impinges on a curved surface, it produces energy, the question arose as to what that energy was used for biologically. QE was capturing particles that transited the semipermeable membrane, so such energy would have been used to stabilize QE, referring to nonlocal gravity. At its origin, that process would have initiated the latticework of life, mediating the assimilation of factors in the environment. Therefore, both that which was inside the cell and that which was in the Cosmos were under the auspices of the same Laws of Nature.

Using the "reverse-evolution" approach to emergence led from Symbiogenesis to Quantum Entanglement/nonlocalization: But where did QM evolve from? Particle Physics, including QCD theory as the common lattice work of the Cosmos

If Symbiogenesis is the "product" of QE/nonlocalization, what gave rise to QE? Prior to the existence of the protocell, there were only particles in the Cosmos. Perhaps they can be accounted for by Lattice Quantum Chromodynamics such as gluons-quarks-antiquarks.

References

Carré, A., Szinnai, G., Castanet, M., Sura-Trueba, S., Tron, E., Broutin-L'Hermite, I., Barat, P., Goizet, C., Lacombe, D., Moutard, M.L., Raybaud, C., Raynaud-Ravni, C., Romana, S., Ythier, H., Léger, J., Polak, M., 2009. Five new TTF1/NKX2.1 mutations in brain-lung-thyroid syndrome: rescue by PAX8 synergism in one case. Hum. Mol. Genet. 18, 2266–2276.

Purevdorj-Gage, B., Sheehan, K.B., Hyman, L.E., 2006. Effects of low-shear modeled microgravity on cell function, gene expression, and phenotype in Saccharomyces cerevisiae. Appl. Environ. Microbiol. 72, 4569–4575.

Sacco, R.G., 2019. The predictability of synchronicity experience: results from a survey of Jungian analysts. Int. J. Psychol. Stud. 11, 46–62.

Sagan, L., 1967. On the origin of mitosing cells. J. Theor. Biol. 14, 225–274.

Saintonge, J., Côté, R., 1984. Fetal brain development in diabetic guinea pigs. Pediatr. Res. 18, 650–653.

Torday, J.S., 2003. Parathyroid hormone-related protein is a gravisensor in lung and bone cell biology. Adv. Space Res. 32, 1569–1576.

Torday, J.S., 2019. The singularity of nature. Prog. Biophys. Mol. Biol. 142, 23–31.

Torday, J.S., Miller Jr., W.B., 2018. Terminal addition in a cellular world. Prog. Biophys. Mol. Biol. 135, 1–10.

Wurm, G., Teiser, J., 2021. Understanding planet formation using microgravity experiments. Nat. Rev. Phys. 3, 405–421.

Afterword

The application of Quantum Mechanics to physiology has interesting ramifications. We know that when tissues undergo injury repair, they will revert backward in their developmental motifs to earlier stages when the endoderm, ectoderm, and most of all the mesoderm were more "plastic." In so doing, the cells involved can re-establish homeostasis, providing the cues for the cell-cell signaling characterizing their normal development in order to repair themselves. If the injury is too severe, the cells will form scar tissue instead. This process is also characteristic of evolution, physiologic stress causing breakdown in homeostasis leading to the production of reactive oxygen species known to cause gene mutations and duplications. In re-establishing their homeostatic set-point under the prevailing environmental conditions, the tissues will evolve with ultimate reference to the First Principles of Physiology within the protocells that gave rise to life. The subsequent symbiogenic assimilation of factors in the environment did so due to Quantum Entanglement, referencing the Quantum Mechanical basis for life, both physical and mental. Thus, our future evolution can be predicted based on these principles.

Author Index

Subject Index

Note: Page numbers followed by *f* indicate figures.

M

N

O

P